FAMILY MATH

Achieving Success in Mathematics

Grace Dávila Coates and Virginia Thompson
with
Karen Mayfield-Ingram and Beverly Braxton

Illustrated by Ann Humphrey Williams
Produced by Louise Lang

FAMILY MATH II: Achieving Success in Mathematics

About EQUALS and FAMILY MATH

Since 1977, EQUALS has developed innovative mathematics curriculum materials to increase access and equity for all students and to help children everywhere realize success in mathematics. We have a special focus on traditionally underserved groups—females, students of color, and those from low-income and language minority families. EQUALS, FAMILY MATH and its Spanish counterpart MATEMATICA PARA LA FAMILIA serve PreK–12 educators, parents, and children. As leaders in the fields of mathematics, equity, and bilingual education, our academic staff present workshops and institutes as well as write books for home and classroom use.

To receive information about EQUALS and FAMILY MATH, to order additional book copies, or to request our publication brochure, contact:

University of California, Berkeley
FAMILY MATH
Lawrence Hall of Science # 5200
Berkeley, CA 94720-5200

510.642.1823 – program
800.897.5036 or 510.642.1910 – books
510.643.5757 – fax
www.lawrencehallofscience.org/equals
equals@uclink.berkeley.edu – program email
eqs_pubs@uclink.berkeley.edu – publications email

Credits:
Production Manager: Louise Lang
Cover and Principal Artist: Ann Humphrey Williams
Senior Editor: Kay Fairwell
Design: Carol Bevilacqua
Photographer: Elizabeth Crews

©2003 The Regents of the University of California. All rights reserved. Printed in the United States of America. Permission to reproduce these materials is granted for home, classroom, and FAMILY MATH workshop use only. For all other purposes, request permission in writing from FAMILY MATH at the Lawrence Hall of Science.
ISBN 0-912511-30-3

Printing (last digit) 10, 9, 8, 7, 6, 5 - 3/2012

Table of Contents

Preface v

Acknowledgments vi

Balloon Ride Revisited 1

Creating an Environment for Learning 2

Questions That Promote Mathematical Thinking 4

How Is My Child Doing in Math? 6

Working With Your Child's Teacher 9

Sample Activity Page 12

Create a FAMILY MATH Kit for Home 14

Probability and Statistics Introduction 16
    Birthday Graph 18
    Sandwiches 20
    Snail Races 22
    Zoo Game 24
    Grandmothers 27
    Lost in Space 28
    Ice Cream Cone Math 30
    About Your Height and More 32
    License Plate Explorations 34

Algebraic and Logical Thinking Introduction 38
    A Penny Saved 40
    Stairway to Seven 42
    No Twin Nim 44
    Circular Nim 46
    Odd and Even Ended Nim 47
    Up the Down Staircase 48
    Baseball Cards 50
    Paths and Ponds 54
    Patios and Paths 57
    Grampa's Coins 58
    Royal Family Puzzle 62

Table of Contents

Number Sense Introduction..64
- This Is Six..66
- Clean Your Plate!...68
- Beaded Braids: Investigating Patterns and Ratios.......................................70
- FAMILY MATH on the Go...74
- Box Math Addition and Subtraction...76
- Consecutive Sums..82
- Bean Boxes..84
- Doubling Bean Boxes...86
- Loopy Lou I...88
- Loopy Lou II..90
- Two Coin Problem..92
- Box Math Multiplication and Division..94
- Family Garden..100
- Family Garden Game...104

Geometry Introduction...106
- Sidewalk Math Using Sidewalk Chalk Introduction..108
    - Line Segments and Intersections...110
    - Me and My Shadow..112
    - Body Trace..114
    - It All Adds Up!...115
    - Value Line..116
- Name Game..118
- Dot to Dot...120
- Line Symmetry..122
- Many Faces of a Cube I...126
- Many Faces of a Cube II..128
- Measurement and Estimation...130
- Spaghetti Geometry...132
- Matting Pictures...136

Organizing a FAMILY MATH Class Series..142

What Mathematics Is Taught in Grades K–6?..160

Masters for Charts, Graphs, and Spinners 174

Other Resources..183

Bibliography...185

Index..186

Preface

We are delighted to present *FAMILY MATH II: Achieving Success in Mathematics*. Since 1977, EQUALS has provided educators and families with challenging and engaging ways to solve mathematics problems. This problem-solving approach has created access to deeper mathematical understanding for teachers, parents, and students.

We believe that mathematics is for everyone. In FAMILY MATH, our special focus is on children and adult family members exploring mathematical ideas. Together, families think about and share problem-solving strategies and learn from one another's process. This all takes place in an environment that is friendly and invites interaction.

Education is about having more freedom to make choices—mathematics makes more choices available to us. The more mathematics we know, the more control we have over higher education and career choices. The more mathematics courses we take, the higher our income potential. People who take more mathematics courses have more options available to them.

In addition to innovative mathematics activities, this book provides tools and ideas to help parents become proactive partners in their children's education.

FAMILY MATH encourages communication, involvement, and playful investigation in mathematics. We hope you and your family will enjoy this book.

The EQUALS and FAMILY MATH Staff

Acknowledgments

The authors of this book are pleased to acknowledge help, support, and inspiration from many sources.

We are grateful to the founders and friends of EQUALS whose vision, commitment, and tireless efforts on behalf of equity in mathematics education continue to inspire our work.

Many thanks to Jean Kerr Stenmark, co-author of the first *FAMILY MATH* book and *FAMILY MATH for Young Children: Comparing*. Her recommendations for this book were invaluable. She continues to inspire honesty, simplicity, and rigor in writing. We also thank Jean's daughter, Jane Stenmark, for her insightful suggestions.

In addition, we would like to express appreciation to Steven Jordan and Mary Burmester. Their contributions to this book reflect their expertise in mathematics, children's learning, and family involvement.

Funding for the initial development and dissemination of the FAMILY MATH program was provided by the Fund for the Improvement of Postsecondary Education of the U.S. Department of Education. The National Science Foundation and the U.S. Department of Education funded further dissemination of the program to Spanish-speaking and English-speaking families.

The Carnegie Corporation of New York provided funding for the publication of the first *FAMILY MATH* book and contributed to the creation of the program and its extensive dissemination throughout the United States and abroad. *FAMILY MATH* activities have been translated into fifteen languages and has served over five million families. The Charles A. Dana Foundation of New York provided funding for the development of *FAMILY MATH—The Middle School Years*.

We thank our FAMILY MATH friends from around the world for their ongoing support. They piloted new activities, provided feedback and encouragement for this book. They are the parent, community, and teacher leaders of FAMILY MATH classes; the national and international site coordinators and directors; and our colleagues in the fields of mathematics, family involvement, equity, research, and mathematics education.

We are grateful to the EQUALS staff. They provided ideas, insights, and encouragement throughout the process of the development of this book. They are José Franco, Director; Terri Belcher, Bob Capune, Ellen Humm, Louise Lang, Deborah Martínez, Helen Raymond, Oralia Ramírez, Claudia Sagastume, and María Villegas.

Most of all we thank the families of FAMILY MATH, who have made this program a part of their lives and helped make it what it is today.

Grace Dávila Coates
Director, FAMILY MATH

Virginia Thompson
Founding Director

Karen Mayfield-Ingram
Contributing Author

Beverly Braxton
Contributing Author

GETTING INTO FAMILY MATH!

Balloon Ride Revisited 2-6
A NIM GAME

The hot air balloon is racing back to town. You can help the balloon clear the top of the mountain by releasing 16 sandbags. However, you can only release one, two, or three sandbags at a time. Release the last sandbag, and save the day!

How

- Play with a partner. Place 16 toothpicks on the game board (facing page) to represent the ropes holding the sandbags.

- Take turns removing 1, 2, or 3 toothpicks. The person who takes the last toothpick(s) saves the hot air balloon.

- Play several times with a family member or friend.

- Work together and discuss strategies for the best ways to play. Play until each person has had a chance to remove the last toothpick(s).

Here's More

- Use fewer or more toothpicks depending on the age of your children.

- Take 1, 2, 3, or 4 toothpicks on a turn.

- Take only 2 or 4 toothpicks on a turn.

- Play with three people instead of just two. How does that change the game?

- For more Nim games see pages 44–47 in this book and also in the original *FAMILY MATH* book.

This is about developing logical thinking and sharpening mental math skills.

MATERIALS
16 toothpicks or markers per two players
Balloon Ride Game Board, page viii

MATH CONNECTION
Nim games are challenging strategy games that help us recognize number patterns. There are many variations of the game from many countries, and there is a winning strategy for each. You will find that some are easier to figure out than others.

REAL-WORLD CONNECTION
In 1994, the Nobel Prize in Economic Sciences was awarded jointly to: John C. Harsanyi, John F. Nash, and Reinhard Selten based on their work in applying game theory to economics. Part of John Nash's life inspired the movie *A Beautiful Mind*.

Thinking through game strategies helps develop the analytical reasoning used by mathematicians and scientists. These techniques can also help in planning and making decisions in everyday life.

Creating an Environment for Learning

The most important thing we can do to create an environment for learning is to love learning. To wonder aloud about the world gives our children the gifts of observation and questioning. Here are some ideas that support children's learning. They are a result of parents and children talking about what worked for them.

"Believe in me."
Believe that our children can succeed, and let them know it.

"Listen to me."
Math is more than numbers. It requires reasoning and logical thinking. When children have more to say or questions to ask, we take the time to really listen. Show respect for a child's ideas, especially if we don't agree.

"Help me do it myself."
Children need guidance, questions, hints, or clues. Resist the temptation to give them the answers to a problem too soon. When we let our children make mistakes, they learn how to solve problems for themselves. This creates a sense of self-reliance and confidence.

"I will solve this."
Model persistence and not giving up. If there is a problem that cannot be solved in the moment, be sure to return to it. Let children know that we are doing so. Let mistakes become learning opportunities by talking about other possibilities for solving a problem. Pose the question, "Is there another way to do this?"

"Hey, look at this!"
Be a keen observer. Point out interesting patterns, bugs, shapes, and events to children. They will develop a sense of observation and curiosity that will help them in figuring out the world around them.

"I am finished with my homework."
Often homework is the cause of stress in families. Look at your child's completed work. Provide helpful feedback. A consistent schedule helps children know what is expected. Treat homework as part of the daily family routine. There is a correlation between success in mathematics and the amount of homework done.

Create a quiet place to study. This looks different for everybody. Let your child help in the planning and decorating.

Questions that Promote Mathematical Thinking

Good news! We need not be math experts in order to help our children succeed in mathematics. Asking questions enables us to lead children to a deeper understanding of mathematical concepts. Good questions open paths to new ideas. This creates an interesting, non-threatening environment that invites learning.

More good news! It is not necessary to know all the answers. Have you noticed that this book (and all EQUALS books) does not include an answer section? We believe that really understanding mathematics comes from a process of investigation—not memorizing one "right" answer. In FAMILY MATH, a child's reasoning and curiosity are exciting ingredients to learning.

FAMILY MATH leaders, educators, and parents developed the following questions as they participated in FAMILY MATH events. These questions do not have a "yes" or "no" answer, or only one possible answer like, "What is two plus one?"

As you ask these questions, listen carefully to your child's responses. Encourage your child to think aloud through the activities, step by step. Be patient when your child is hesitant or confused. Let your child complete the activities and make mistakes without making judgments. Praise your child's thinking. Be sensitive to when your child feels confident and is ready to move on.

The following questions will help keep the discussion going as you and your child explore mathematics together. You may want to add some questions of your own.

What ideas do you have about how we might begin?

Are there other possibilities? If so, what might they be?

Can we make a model? Describe it. Build it.

What other similar problem have you done?

How can we be sure about that?

Hmm, I had not thought of that. Tell me more about it.

How can we check to see how close your guess is?

How can we do this differently?

How did you decide which objects go in the circle?

How did you figure that out?

Is there a pattern here? Describe it.

Is there anything missing? How do we know?

Tell me about your design.

Tell me how you did that.

What do you suppose would happen if…?

What is most (least) likely?

What other things can we find shaped like a square/circle/triangle…?

What other ways can we show that?

What will you do next?

Why is this useful?

Why do you think that?

Why does this work?

Be careful not to ask too many questions all at once. One or two well-placed questions encourage thinking and help create deeper understanding.

How is My Child Doing in Math?

How do we know how well our children are doing in math? Do standardized tests tell us enough? What effect do standardized test scores have on students? Are there more meaningful ways to assess how a child is progressing and the mathematical understanding that is taking place?

There are many kinds of tests in school. There are ongoing classroom tests such as spelling, mathematics, and reading tests. They tell us what information a child has memorized or retained in these subjects.

Standardized tests are ones that children take once a year, and the scores from these tests may follow our children for their entire school lives. Standardized test scores tell us very little about how well our children understand mathematics. Also, they do not provide a classroom teacher with the detailed information necessary to improve a child's skills.

One use of standardized tests is the placement of students in specific math classes based solely on their scores. Misuse of these scores demands our attention as our children progress from the elementary grades to middle and high school. There are things you can do early on to act as advocates for your children when important decisions are being made.

Assessment that helps students learn should inform us and guide teaching. Although standardized tests have been widely used for the last fifty or sixty years, there are more revealing ways to assess children's progress. Portfolios, math journals, and performance assessments are such examples.

In the following assessment models, you will find ideas for positive actions. Ask your child's teacher if these are a part of the mathematics program in your child's classroom or school.

Portfolios

A portfolio is a collection of a student's work. A portfolio may include answers to open-ended questions, videotapes, audiotapes, records of performances, art, or other models. To assess a portfolio we might consider:
- The students' understanding of the mathematics they are studying
- Extension of the work beyond what has been taught
- Clarity in presenting ideas
- Growth in social and academic skills that may not be reflected by a standardized test
- Success in meeting the criteria determined ahead of time by teacher and students

Math Journal

Mathematics writing may take many forms. A math journal allows us to see the progression of a child's thinking and understanding. Interactive journals are those in which the teacher responds to the student's entries. This is an excellent model.

As an assessment tool, it is important that children be provided with feedback and be allowed to revise their work (a common process in writing classes). The revised version becomes the official paper for evaluation. Students may write about a mathematical investigation or an open-ended problem.

Performance Assessment

When we observe children solving problems, completing tasks, and generating ideas, we get important information about their level of understanding. We can see how a child reasons (thinks logically) and how that knowledge is communicated to others. Through performance assessment we can observe how a child:
- Persists and works independently or cooperatively with others
- Observes and makes hypotheses
- Thinks flexibly and changes strategies when one way doesn't work
- Finds patterns and uses them in solving problems
- Collects, organizes, and displays information
- Finds resources to complete a task or solve a problem
- Communicates using mathematical language

How is My Child Doing in Math?

Things You Can Do

We mentioned portfolios, math journals, and performance assessments; and there are others you may wish to explore. Establishing open communication with a child's teacher is an important first step. You can also:

- Observe the classroom. Observe students interacting with one another. Are they asking questions and posing new ideas?

- Involve children in their own assessment. Review mathematics they did in the fall or the previous year and compare with current work. Ask them to describe the improvements they see.

- Ask questions that require critical thinking when helping your child with homework. For example you might ask, "Which of these would work best? Why?"

- Create a portfolio for your child at home. Add work that both you and your child choose. Add samples over time so you can see progress.

- Ask your child's teacher about the tests or assessments that are given in class. What is the purpose of the test? How will my child be affected by the results? What can I do to prepare my child for success with this assessment?

The ideas we listed are just a beginning. Assessment that is meaningful engages children, informs teachers and parents, and results in a more complete picture of what a child really understands.

Please read Working With Your Child's Teacher on page 9 and Questions That Promote Mathematics Thinking on page 4.

Working With Your Child's Teacher

Do you remember your parents' interactions with your teachers? Were your parents informed in person about your progress or only by report cards? Were they given school rules so that you might follow them? Were they called only when you were in trouble? Did a teacher ever call your parents just to tell them how well you were doing in a subject?

We know that children whose parents are involved in their schools experience higher academic success than those whose parents are not. Children learn best when their parents and teachers work together. This is why it is important that parents have a good working relationship with their child's teachers.

Often teachers reach out first, but we can share the responsibility for creating and maintaining strong home-to-school communication. The following strategies can help you get to know your child's teacher and get the teacher to best know your child.

- Initiate communication in a consistent and proactive (as opposed to reactive) manner. This enables you to stay better informed about your child's progress, learn how to best help with class work or projects, and create a more effective partnership with the teacher.

- Introduce yourself in person at the start of the school year. Let the teacher know how you would like to assist in the classroom, when it is possible, and that you welcome communication on a regular basis. Indicate whether you prefer to be contacted by telephone, email, or by letter. Ask the teacher how she prefers to be contacted.

- Be involved in the school as much as you can be. If you work outside the home, let the school know how you can help in other ways.

- Find out how the school operates. Is there a principal? Do important school policies get decided by the school staff? How are parent voices included in the decision-making process? What enrichment or after-school programs are in place? What are the academic standards?

- Send notes to the teacher. When your child is excited about a project, lesson, or field trip, let the teacher know.

Working With Your Child's Teacher

- When you disagree with your child's teacher, it is important that you first speak to the teacher to clarify the issues and your understanding of them. If you cannot reach a satisfactory solution, then follow up with a conference with other school personnel. Avoid criticizing teachers in front of your child. It may cause your child to become confused or defiant. Children need to see problem-solving models so they may become better at their interactions with others.

Remember—involved parents make the best advocates for their children.

Sample Activity Page

K–6 — Grade Levels
Indicates appropriate grade levels.

MATERIALS
A list of materials you will need for this activity.

MATH CONNECTION
Definitions or explanations of the mathematics demonstrated in this activity may be found here. Other related concepts or ideas will also be included.

REAL-WORLD CONNECTION
This tells us how the mathematics in this activity connects with math-related careers, math in nature, and math in our daily lives.

How
- Gives step-by-step directions for doing the activity.

- There are some game boards and charts that you will need to duplicate. Avoid using the original if you are going to write on it. Copy game boards on card stock or heavy paper. You can also glue paper game boards onto heavy cardboard (like cereal boxes) for durability.

Here's More
These notes provide suggestions for extending and varying the activity. Sometimes we make it more complex or challenging. Other times we might include adaptations for younger children.

This is about the mathematics you will use in this activity.

SIDEBARS
Information in sidebars tells more about the mathematics involved in the activity or provides connections to other topics.

FAMILY MATH II: ACHIEVING SUCCESS IN MATHEMATICS

Create a FAMILY MATH Kit for Home

The activities in this book do not require expensive materials. We believe that using simple items found in the home will provide children with rich mathematical experiences. Buttons, toothpicks, and paper squares are some items you will need for your FAMILY MATH kit. Keep them in the same place so your child knows where to find them. These materials may also be helpful for other homework projects.

Tools you will need in a home FAMILY MATH kit

- scissors, glue, and tape
- paper: plain, graph, and scratch paper
  Be creative and use the blank side of letters, the inside of cereal boxes, or recycled printer paper.
- pencils: color, regular, and thin-tipped pencils, mechanical pencils, and a sharpener
- ruler, straight edge, protractor, and a compass
- calculator
- things to count like beans, buttons, bottle caps, toothpicks, and wooden cubes
- dice and playing cards for games
- string, yarn, and sidewalk chalk

If you do not have all of these things in your home, ask the teacher or school staff if they provide these tools on loan.

One mother had a Class Shower, in this case with a mathematics theme, for her child's teacher. This mother sent out a list of materials needed for the classroom. The class created homework packs in zip-lock bags for children to borrow.

A note about calculators
Calculator use is often a point of debate among primary educators and some parents. The worry is that children will become dependent on calculators rather than memorize the basic facts. We feel that a calculator is one of many tools used in solving math problems.

Should calculators be available to children at all times? Yes, children should become fluent in the use of all tools that help in problem solving. Other tools include a straight edge, a compass, a protractor, and a computer. These are all tools that allow us to solve problems, research information, and discover new ideas.

"FAMILY MATH does not take energy; it creates energy."
—a FAMILY MATH parent

INTRODUCTION

Probability and Statistics

This section is about probability and statistics, two important branches of mathematics that help us understand the world around us. They are often paired because they are interrelated. Have you ever bought a raffle ticket? If so, there was a certain probability or chance that you would win which depended on how many tickets were sold. If you are one of the "one in six" families in the United States who has a video game in their home, you are a statistic.

Probability

Has your child ever flipped a coin to determine who would go first in a game? Have you ever been caught in the rain without an umbrella because you thought the probability of rain was unlikely? These are just a couple of the many examples in which probability or chance plays a role in our daily lives.

Probability is the chance of something happening. When you throw a fair die, the probability that it will land on 2 is 1 chance in 6. We know this because there are six sides on a die and only one side has two dots. But what really happens when you roll the die? If you roll the die six times, you will find that the numbers 1 through 6 rarely come up one time each. There is a chance that the same number will come up two or three times and others not at all. But if you make many rolls, the number of times each number (1 through 6) comes up will begin to even out. Recording the outcome of each throw brings us to statistics.

Statistics

Statistics is the area of mathematics in which we collect and analyze information (data). How do we really know that "three out of four" dogs prefer X brand of doggie treats? Did you know that the average American child watches five hours of television daily? Collecting data and organizing it to gather useful information is what statisticians do.

Statistics have many uses. They can be used to:
- inform businesses (which neighborhoods would support a new coffee shop)
- influence people (more families prefer X detergent)
- create a sense of inclusion (nine out of ten teens wear brand Y jeans)
- predict future behaviors (by the year 2010 more people will drive electric cars)
- evaluate new drugs and medical treatments (the number of people who responded to a new drug was much greater than those who responded to a placebo)

- connect actions to outcomes (if you volunteer at school, your child will do better than if you don't)

We included activities in this chapter that develop an understanding of basic concepts in probability and statistics. This section focuses on the following:
- organizing data or information in graphs, charts, or other visual forms
- making predictions
- analyzing
- thinking logically
- exploring probability through games of chance

As you and your child explore these activities, we encourage you to look for and share real-world graphs and tables. They can be found in newspapers and magazines. As your family discovers the excitement of probability and statistics, you will be joining scientists, business people, and lots of others who are curious about the world.

"It is not knowledge but the act of learning, not possession but the act of getting there which grants the greatest enjoyment."

–Karl Freidrich Gates, Letter to Boylai
Young Mathematicians at Work, page 15

Birthday Graph

K–6

MATERIALS
construction paper
paper squares or Post-it notes
masking tape and pens

REAL-WORLD CONNECTION
What is the likelihood that someone in your neighborhood has the same birthday you do? Birthday Graph illustrates how using math to explore probability can have surprising results.

This activity requires you to keep a graph posted over a period of time so that family and friends can "sign in" when visiting your home.

This is about
exploring probability (chance), creating graphs, and keeping track of our family and friends' birthdays.

1-10											
1-6		3-15						9-9			12-12
1-3		3-6				7-9		8-4		11-22	12-22
JAN	FEB	MAR	APR	MAY	JUNE	JULY	AUG	SEPT	OCT	NOV	DEC

How

- Create a graph similar to the above illustration, and post it on your refrigerator or a space that is easily accessible and visible.

- Have all members of your family write their names and month and day of birth on the graph above their birthday month.

- Your "birthday twin" is another person whose birthday falls on the same month and day as yours. When extended family, friends, and neighbors visit, have them add their names and birth dates (using month and day only) to the graph. Estimate how many people will be on the graph when the first "birthday twins" appear who are not fraternal or identical twins.

- How many people had signed up on the graph when the first "birthday twins" appeared?

- Ask your child to "read" the graph. Pose some questions, such as:

    How many more people have birthdays in May than in December?

    Which months have zero birthdays listed?

    If two more birthdays were added to the month of ___, how many would that make?

Asking questions helps you see how much your child understands and allows for more thinking about graphs.

There are 365 days in a year. You may be surprised to find at least one set of birthday twins on your graph, especially if there are only 25 or 30 people who signed it. Mathematicians can calculate that it takes only 23 people for there to be a 50% chance of a double birthday. This does not mean that you will have a double birthday on your graph when the 23rd person signs. It means that if a lot of families in your neighborhood make birthday graphs, the average number of people on the graph when a double birthday appears will be 23.

Sandwiches

This is about
creating combinations where the order of the items chosen does not matter.

MATERIALS
one set of Sandwich Cards, page 21

Being able to identify combinations is very important in probability theory. On the other hand, in Ice Cream Cone Math and part of the Licensing Math activities, the order you choose does matter. These activities—when the order <u>does</u> count—deal with permutations.

Very young children will enjoy making the different combinations and comparing them. Fourth graders and up should be able to make all the different combinations and predict or find their total.

Counting combinations is part of a field called discrete math. Discrete math includes counting, finding the optimal path or route, and finding the best strategy in a game.

How

- The Calliope Cafe offers a sandwich special. You have a choice of Swiss cheese, sliced turkey, or tuna salad fillings with tomatoes or lettuce on wheat or white bread.

- Copy the sandwich cards on page 21 and cut them out.

- Move the sandwich cards around to help you figure out how many different sandwiches can be made using one filling, one of the vegetables, and one type of bread.

- How many possibilities are there if you use two fillings in each sandwich?

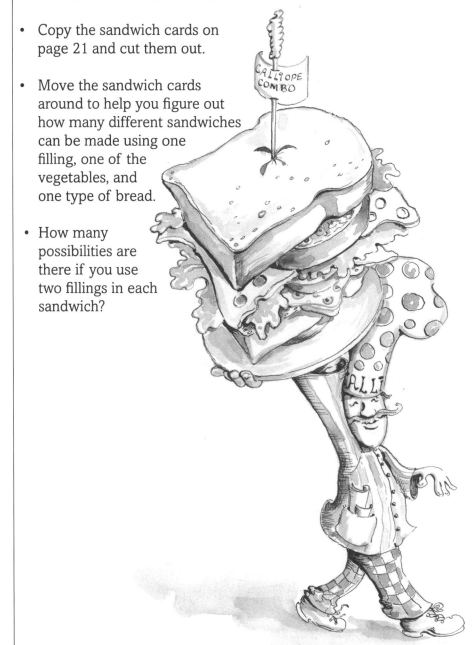

SANDWICH CARDS

WHITE BREAD

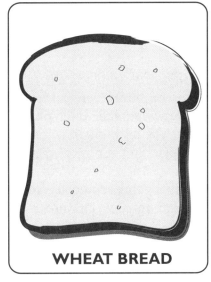
WHEAT BREAD

TUNA SALAD

WHITE BREAD

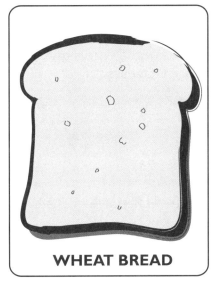
WHEAT BREAD

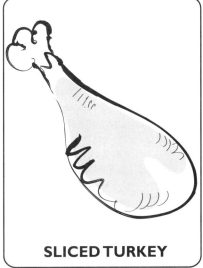
SLICED TURKEY

LETTUCE

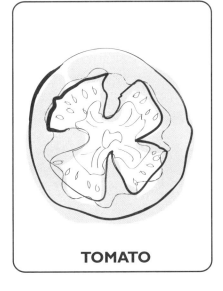
TOMATO

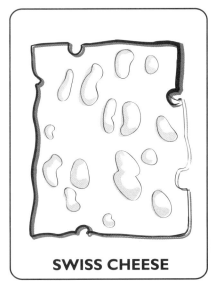
SWISS CHEESE

Snail Races

This is about
playing with probability as we develop number sense.

MATERIALS
crayons
game board, page 23
40 beans or markers (small enough to fit in the game board circles)
two dice

REAL-WORLD CONNECTION
Sometimes we can predict events based on past experiences. For example, if it rained in your town every first week of December for the last thirty years, one might say it is almost certain that it will rain at the same time this year. We explore probability based on the outcomes of rolling two dice. This means that regardless of past experience, one cannot predict which two dice will come up next.

How

- One to four people can play this game. Use the game board on the facing page. Each player must choose a snail they think will reach the garden first. Identify the snail as yours by coloring it or by placing your initials on it.

- In this game, you will find yourself advancing any of the snails including your opponents' snails.

- The first player rolls two dice and adds the numbers together. Place a marker on the first circle of the snail with that number.

- Keep taking turns rolling the dice. To advance a snail, place another marker on the next circle each time that snail's number comes up. This means that every circle in a snail's path is covered as it advances.

- Remember all the snails are racing. That is to say, you may have to place markers on a snail that is not your own.

- Keep rolling the dice until one snail crosses the finish line and reaches the garden for a tasty snack.

- Play at least four times. Keep track of the races you played and the snails that won.

- Which snails moved at a snail's pace?

- Which snails won the most? Why do you think this happened? Discuss any patterns that you and your family observed.

Zoo Game

K–6

MATERIALS
dice
markers
game board, page 26
scratch paper and pencils

MATH CONNECTION
Probability is sometimes called the science of chance. What is the probability that a particular event will occur? We can calculate probabilities mathematically by looking at all of the possible outcomes of an event. For instance, a coin has 2 sides, usually called heads or tails. When it is tossed in the air, the probability of a fair coin landing on either heads or tails is 1/2, meaning one out of the two possible outcomes or 50%. Probability is at work when you play games that use dice or spinners.

You and a friend have arrived at the entrance to the zoo, and both of you are trying to decide which animal to visit first. You want to visit every major animal exhibit. To get to any exhibit you must move along a system of paths and plazas. From each plaza you can choose between a left- or right-hand path.

To determine where to go, you and your friend decide to use a die and make your visit into a probability game.

When each of you reaches an animal exhibit, you go back to the entrance and start again.

How

- Play in pairs. You will need one marker for each player. Share one die and the game board. Decide who will go first.

- Predict which animal exhibit you will visit the most.

- Place your markers at the zoo entrance. Take turns rolling the die. It is important that you move the game board to face the person rolling the die. In this way, players can easily read the game board as they move their markers.

- If the die shows an even number, move your marker along the path to your right until you reach Plaza 2. If the die shows an odd number, move your marker along the path to your left until you reach Plaza 1.

- Continue taking turns rolling the die. When you roll an even number, move your marker to the plaza on your right. When you roll an odd number, move your marker to the plaza on your left.

This is about developing an understanding of probability using odd and even numbers.

- When you have played a few rounds each, talk about how many more rounds you think it will take to visit all the animals. When you have visited half the animals, look at your original prediction. Would you like to revise your prediction?

- Play this game several times, and compare your results with other players. Were your results the same? Which exhibits were you most likely to visit? Least likely? Try organizing your information in a different way.

- Investigate what you think is happening in this game.

Here's More
- Suppose you want to see the Sun Bear most of all. Predict how many rounds it will take to get to the Sun Bear exhibit. Play until you get to the Sun Bear. How close was your prediction?

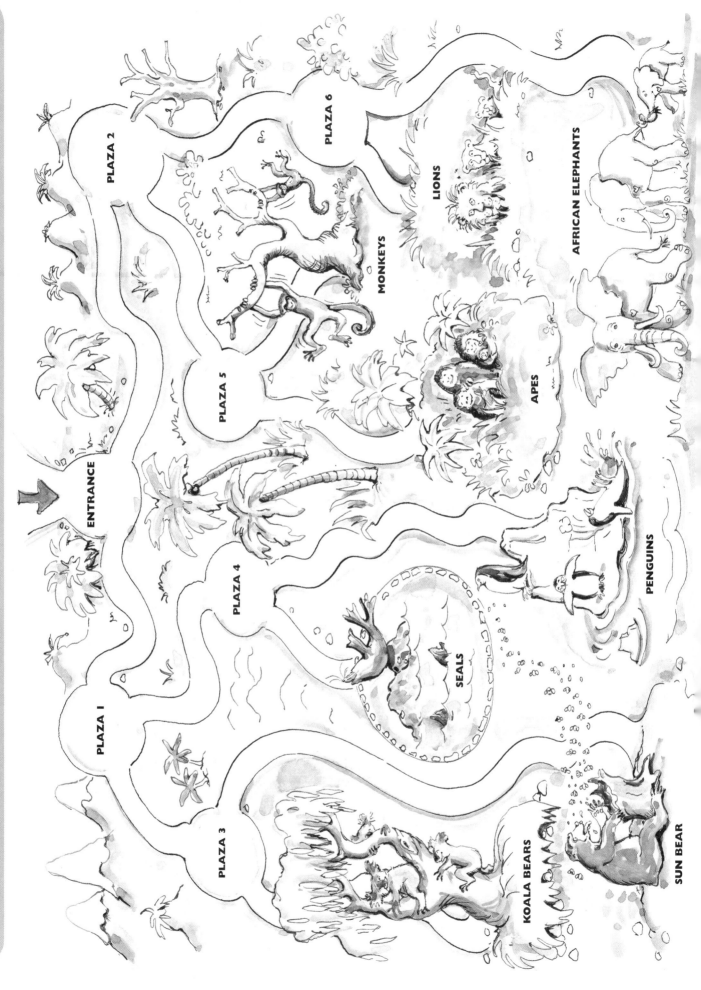

Grandmothers
K–6

Jenny is trying to figure out how many biological great grandmothers she has. Whenever she gets to Granny Rose, she loses count. Draw a picture to help Jenny out.

How

- Draw a picture to find out how many biological great-grandmothers Jenny has. If you like, you can use markers to represent each of the grandmothers.
  Hint: First figure out how many grandmothers she has.

- How many great-great-grandmothers does she have? Continue your picture to help figure it out.

- Discuss with your family how you worked on the problem and what you found out. What would your family tree look like?

Here's More

- Make a Fraction Kit, and try the Fraction Kit Games activities from the original *FAMILY MATH* book. How can these activities help you with Grandmothers?

This is about
counting, developing mathematical reasoning, and creating diagrams to explore exponential growth.

MATERIALS
scratch paper
pencils
markers

MATH CONNECTION
Discrete mathematics involves finding efficient ways to count separate things. In this activity the separate things are grandmothers, great-grandmothers, and great-great-grandmothers.

Lost in Space
K-6

This is about getting experience with probability, number sense, and direction.

MATERIALS
game board, page 29
two cubes (1" by 1")
marking pens
markers such as beans

You and your partner just completed repairs on the outside of the Spaceship Falcon. It's time for lunch, and the Intergalactic Pizza Express just delivered a delicious Galactic Combo Pizza with double cheese. Get your slice of pizza by rolling the right combination of numbers and letter.

How

- Mark the first cube with 3 F's (for the Spaceship Falcon) and 3 S's (for outer space).

- On the second cube, mark two sides with the number 1, two sides with the number 2, and two sides with the number 3. These numbers represent 1 to 3 steps in the air passage toward the safety of the Spaceship Falcon (and a great lunch) or toward outer space.

- Each player places a marker on Enter on the Astro Vac.

- Players take turns rolling both dice. One die will indicate which direction to travel, either toward the spaceship or toward outer space. The other will indicate how many spaces to move.

- Each move to the left takes the player towards the Spaceship Falcon; each step to the right takes the player toward space and no pizza.

- A player can only enter the Spaceship Falcon by rolling the exact number needed to set foot inside. If a player falls into outer space, that player must start over.

- Play four games. How would the game be different if there were 11 sections of the air passage instead of 9?

- Talk about what would happen if there were 4 F's and 2 S's on one die? How does the game change if there are more 1's than 3's on the second die?

28 FAMILY MATH II: ACHIEVING SUCCESS IN MATHEMATICS   PROBABILITY AND STATISTICS

LOST IN SPACE GAMEBOARD

Ice Cream Cone Math

3–6

MATERIALS

ice cream cones and scoops, page 180
crayons or markers
scissors
construction paper of different colors (optional)

Discrete math involves finding ways to count things. A permutation is an arrangement of a set of objects in a particular order. We count the objects, and we count the number of arrangements that can be made with those objects.

Mathematicians would say that we are finding permutations here because we care about the order. For example, the arrangements of colors that follow are some of the different permutations of red, white and blue:

>   red, white, blue
>   red, blue, white
>   blue, red, white

Sylvia will eat only triple-decker ice cream cones made with three *different* flavors. She considers a *different* order of the same flavors to be a "different" cone. Sylvia has chocolate, vanilla, and mango ice cream in the freezer.

How many different cones can Sylvia eat before she has to repeat an arrangement (permutation)?

How

- To begin, create at least 12 cones and six scoops of each flavor. Color each flavor a different color. Use the cones and scoops on page 180, or make your own using construction paper.

- Use the cones and scoops to make all the possible arrangements that Sylvia can create. Illustrate or make a list of all of the possibilities.

- Work together to find other ways to organize your results. Explain how you know you found all of the arrangements.

- How many different triple-decker cones are possible if Sylvia added toasted almond and had four flavors of ice cream instead of three? You may need to make more cones and scoops to work with.

- What will happen when a fifth flavor is added? Do you see a pattern?

- Model thinking aloud. As you create the ice cream cones with your child, talk about your strategies. As your child develops confidence with the ice cream scoops, try creating permutations with items your child routinely uses such as crayons, barrettes, beads, or stickers. Let your child select favorite items. Young children can explore freely and talk about the different permutations they find.

The number of permutations with three different flavors is 6; it can be thought of as 3 x 2 x 1 or three choices for the first flavor times two choices for the second flavor times one choice for the last flavor. Mathematicians have a short cut. They write 3 x 2 x 1 as 3!, which is read: three factorial.

Here's More

- How many permutations would there be if Sylvia ate quadruple-decker cones with all five different flavors? Try using the factorial system to help you figure this out.

- Sammy also eats only triple-decker ice cream cones. He also cares about the order of the scoops, but doesn't mind two scoops that are the same flavor as long as the other is different. Using chocolate, vanilla, and mango, how many different ice cream cones are available to Sammy?
  Hint: Start with only double-decker cones. You may need to make more scoops and cones.

- What happens if Sammy eats quadruple-decker cones with these three flavors? As you do this, try to use what you have learned without using more cones and scoops.

- Try Sandwiches and License Plate Explorations.

About Your Height and More

3-6

MATERIALS
pens
graphs
non-stretchy string
scissors

REAL-WORLD CONNECTION

A ratio can be expressed as:
- a probability, like flipping a coin
- a percent—70% chance of rain
- a speed—55 miles an hour
- an average-average income, taking the total salaries divided by the number of people

Collecting data and examining it for trends is important in the field of statistics. This kind of procedure is used in all experimental sciences and many social sciences such as business and economics.

This is about measuring, estimating, and organizing data to understand ratios.

This activity invites us to explore ratio by looking at how our height compares with different body measurements.

In About Your Height and More, when you compare the length of your body height to the number of times that length will circle your fist, foot, and so on, you are making a comparison of 2 parts. After the data is recorded for each person, look for common ratios for different pairs of measurements.

How

- Ask a friend or family member to help you cut a string that measures your height.

- Keep a record of everyone's answers to the questions below. Use one color pen for adults and a different color for children.

    1. How many times does the string measuring your height go around your head? Be sure to write your own answers as a ratio. See *About Ratio's* for two ways to do this.

    2. How many times does the string go around your wrist?

    3. How many times does the string equal your foot's length?

- What do you notice? Are these ratios true for everybody? Discuss your reasoning.

Here's More
- Cut a string that is the length of your foot. Estimate how many times the string you cut will go around your fist.

- How close was your estimate? Compare your measurement with other family members' measurements.

- Can you make any general statements about body measurements and their relationships to each other?

Note: Did you know that some people shop for socks using their fist as a measurement for their sock length? Discuss how you might use the information from your foot and fist measurements when shopping for socks.

ABOUT RATIOS

A ratio is a way of comparing two values. The ratio of any value a to any value b can be written either in fraction form that is, as $\frac{a}{b}$ or in the form a:b.

License Plate Explorations

4–6

MATERIALS
paper and pencils
3 sets of each letter of the alphabet, pages 36–37
numbers 0–9, page 37

REAL-WORLD CONNECTION
There are many things that are made up of numbers in our lives, such as telephone numbers, license plates, and zip codes. The designers of these systems need to plan ahead to make sure that there will be enough numbers available as the population grows. To do so, they need to understand the mathematics of finding combinations and permutations.

How

You may wish to try Ice Cream Cone Math on page 30 before trying this activity.

- Copy and cut out four sets of the letters and numbers on pages 36-37.

- Current California license plates for autos have one number followed by 3 letters followed by three more numbers. Let's create license plates on a smaller scale. Suppose you could use only two letters: A and B. Working together, use the letters to find out how many different two-letter sets or permutations you can make. Remember it's OK to repeat letters. Record as you go. When your child shows confidence with two-letter sets, you are ready to move on.

- What happens if you can choose from A, B, and C to make three-letter sets? What different ones can you find now? Look for patterns.

- What happens when D is added? Remember you are making 3-letter sets.

- What other ways are there to keep track of your investigations?

- Talk with your child about other items that can be arranged with as many repeats as are possible. For example, choosing from buttons, bottle caps, and toothpicks, you can line up a button, a bottle cap, and another bottle cap and make a record of that order.

This is about making permutations with repeats, using letters and numbers.

- Choosing from the same items, ask your child to create a different arrangement. Or, you can reverse the roles and you follow your child's example. Following your child's directions will give you an idea of how well your child understands how permutations work.

- Try combining any two numbers (0–9) with any two letters.

- How many different combinations can you find?

Here's More
- What happens if you cannot repeat letters or numbers?

- Try combining any three numbers (0–9) with any three letters.

- Is there a shortcut for finding the different arrangements of items? How did you discover it?

- Encourage younger children to work with only two of their favorite letters or numbers. Keep a record as you go.

- Compare license plates from other states or countries.

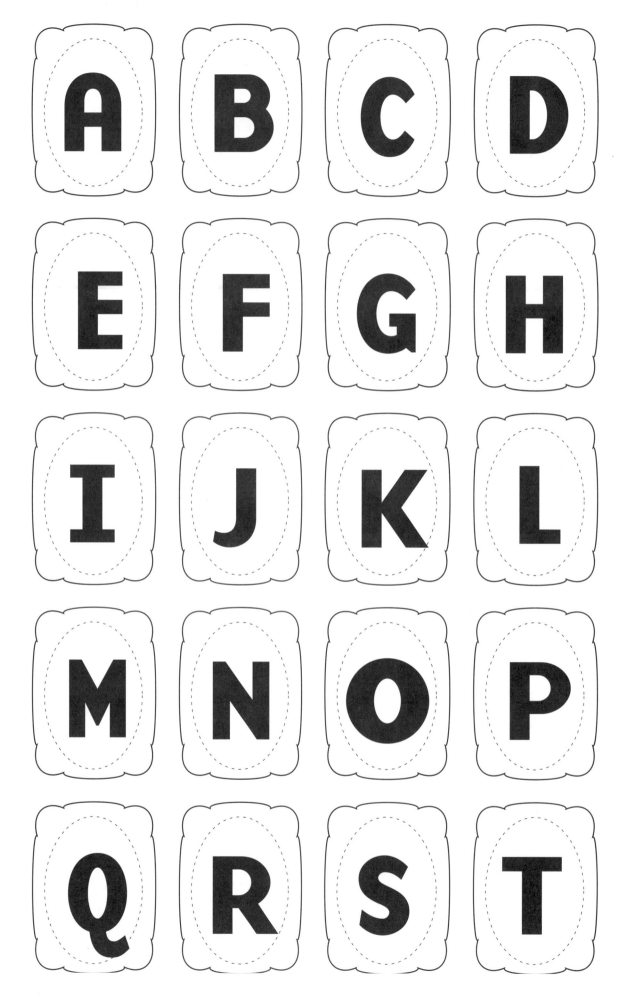

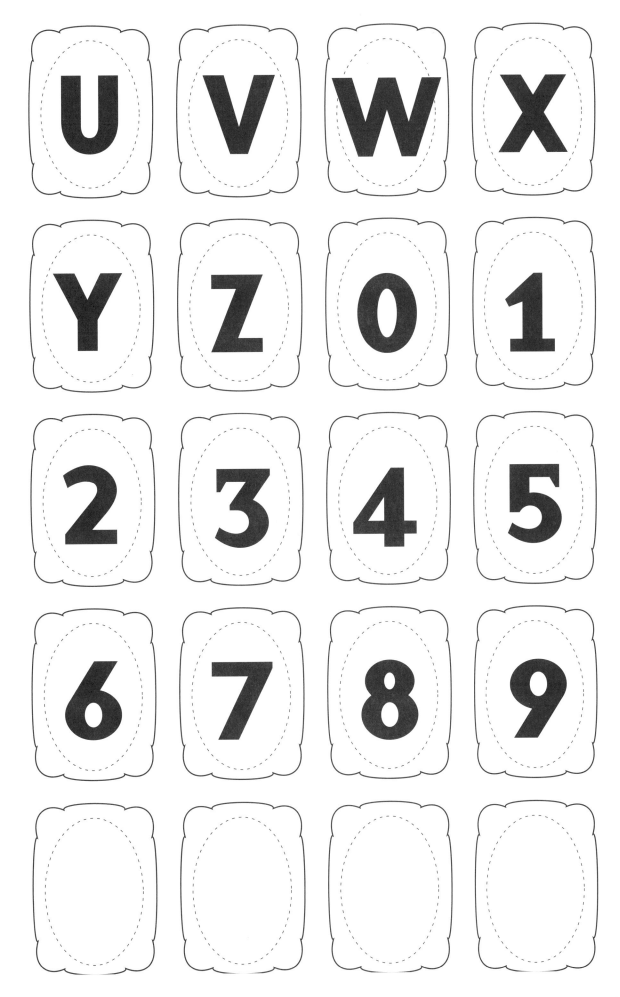

LICENSE PLATE EXPLORATIONS

FAMILY MATH II: ACHIEVING SUCCESS IN MATHEMATICS 37

INTRODUCTION
Algebraic and Logical Thinking

Guess what? We are all algebraic thinkers, and most of us don't even know it! Do you ever think about the rate at which you earn a raise at work? Are you a saver or a spender? If you are a saver, how fast will your money grow? Have you wondered when the right time is to refinance your home? If you do refinance, should you pay points or not? We can use algebra to figure these things out. Algebra is often the target of negative attitudes because many of us think that only "smart" people get it. The truth is, we can all "get it."

This section provides some important activities and Nim games to help your family explore algebra and logical thinking in an informal way. Young children can experience algebraic ideas by using simple things like beans, buttons, or blocks to make models.

Nim games continue to be popular because they challenge players to improve their game every time they play. As children become more sophisticated in their thinking, they develop more effective strategies. Although one can win at Nim by chance, it is logic that is needed to ensure a winning strategy.

The activities in this chapter encourage us to use the language of algebra in ways that relate to our daily lives and experiences. Young children have a strong sense of what is fair. When they say, "He has twice as much as I do!" they are using algebraic language. When they want to figure out how many weeks' allowance (if they get one) it will take to save up for something special, they can use algebra to figure it out.

In middle school, algebra becomes more abstract and formal. At this time several things can happen. Your child may be enrolled in an algebra class because of high test scores and grades in math. Children may be tracked out of algebra due to low math test scores or grades, or be required to enroll in it regardless of how well prepared they are. This is significant because algebra is more often the course that determines whether students will be placed in college preparatory courses. And, while we do not propose that everyone must go to college, we absolutely do believe that everyone should have the opportunity to choose.

Here are some simple things you can do as your family tries out the activities in this chapter:

- Think aloud and share ideas with your child as you try to solve a problem. Using language to describe what is happening helps us develop our problem-solving skills.

- Encourage everyone to organize their work into graphs, tables, or other models so that you can keep track of information. We have provided some examples throughout the activities.

- Make connections to your own experiences; ask your child to do the same. Making these connections helps us to better understand and remember new learning.

- Be curious. If you don't understand an idea or concept, let your child know that you will look for the resources you need to find a solution.

We can all be successful in algebra. We can all be resourceful. High-achieving students are not necessarily the ones that figure things out immediately. They are the ones who know how to keep on trying. They ask questions when they don't understand. When they can't figure it out, they find resources that can help them. When we model this for our children we give a gift that will serve them well throughout their lives.

"A mathematician, like a painter or a poet, is a maker of patterns."
—Godfrey H. Hardy, A Mathematician's Apology
Young Mathematicians at Work, page 1

A Penny Saved

K-3

MATERIALS
100 pennies
pencils
paper

MATH CONNECTION
Most of us were taught to count by twos, fives, and tens. As we began to multiply, we learned to count by threes, fours, and other numbers. Here, we play with numbers to look at number patterns in an organized fashion. This process of looking at patterns of numbers is important in all areas of mathematics, especially in algebra. In arithmetic we work with specific numbers like 1, 2, 3,…; in algebra we use letters to designate general numbers.

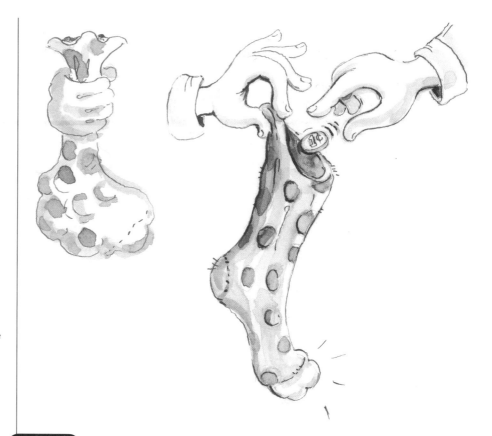

How

- Work together and estimate how much money each of you would have if you saved two pennies for each year of your age.

- Now ask your child to arrange the pennies in twos and count them by twos until the correct age is reached.

- How much money would you have if you saved three pennies for each year of your age?

- How could you organize this information so that it shows others how much money you saved?

PERSON'S AGE	PENNIES SAVED
1	3
2	6
3	9
4	12

- Now, without actually taking more coins, predict how much money you would have 25 years from now if you kept saving three pennies a year. Don't forget to include the pennies you already have.

- Have your child try this activity using different coins like nickels, dimes, and quarters, and predict how the amounts will change for the various coins.

Here's More
- Make some charts to compare amounts with the different coins.

- How many months would it take you to save $100.00 if you were saving 4 quarters per week? 6 quarters per week?

Stairway to Seven
K-6

MATERIALS
wooden cubes
paper
pencil

MATH CONNECTION
This activity provides a way to explore algebraic thinking with a hands-on approach. Building patterns, predicting, and organizing information help us to see how numbers and patterns are related.

Lin wants to build a tree house. His mom said it is OK as long he builds safe stairs that do not go higher than seven steps. The only materials that Lin has are a lot of square wooden boxes his Auntie Helen gave him. How many boxes does Lin need to build the stairs?

How

- Arrange some of the boxes so they form stairs that go up seven steps:

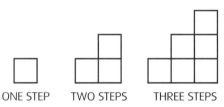

ONE STEP TWO STEPS THREE STEPS

- When you have built the third step, ask your child to predict how many boxes Lin will need to build the seven steps. If it is too soon for your child to do this, keep adding steps until a prediction can be made.

- Continue until there are seven steps.

- How many boxes would Lin need if his mom let him build stairs with 10 steps? 20 steps?

- Organize your information to keep track of boxes and steps.

- Here is one way you can keep track:

STEPS	BOXES
1	1
2	3
3	6

- What is a pattern that shows how the number of boxes grows each time you add a step?

This is about predicting, counting, and creating patterns as we explore algebraic thinking.

Here's More
- See if you can predict the number of wooden boxes it would take for Lin to build the 100th step.

No Twin Nim

K–6

MATERIALS
25 markers, buttons, or toothpicks per two players

MATH CONNECTION
There are many variations of the game of Nim from many countries. There is a winning strategy for each. Some are easier to figure out than others.

In No Twin Nim, we look for general patterns when we have a number (call it n) of markers and can pick up 1, 2, up to k markers per move. Mathematicians call this the general case. They would want to know if it matters if you start with an odd number of markers as opposed to an even number. Note that n and k are variables and can stand for any number.

How

- Lay out a row of 12 markers. Play with a partner.

- Take turns removing 1, 2, or 3 markers from the row.

- You may not take the same number of markers removed by your partner on the previous move.

- The winner is the one who takes the last marker(s) or leaves the other player unable to move.

- Play several games with your family. Then discuss your ideas about the way the game works. Talk together and help each other work out a winning strategy for every player.

Here's More

- Play with more people.

- Play with more markers.

- Allow players to take 1, 2, 3, or 4 markers with the same restrictions.

- Change the rules so that you lose when you take the last marker.

This is about creating strategies and developing logical thinking.

Circular Nim

K-6

This is about
practicing strategies and developing logical thinking.

MATERIALS
13 markers, buttons, or beans

How

- Arrange 13 markers in a circle.

- Play with a partner. Take turns removing 1 or 2 markers from the circle. If you take away two markers, they must be adjacent (directly next to each other).

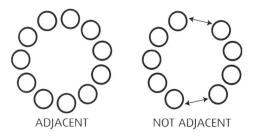

ADJACENT   NOT ADJACENT

- The player who takes the last one or two markers wins.

- Play several games with your partner and then discuss what you learned about the game.

- When you and your child see patterns, work together to figure out winning strategies.

- Here is a question that might help you determine how well your child understands what happens in this game:
    What can you do to make sure that you take the last marker(s)?

Here's More

- Try playing with 16 and 21 markers in a circle. Cooperate to develop the best winning strategy.

- How does your strategy change when you begin with odd or even numbers?

- Allow players to take 1, 2, or 3 markers with the same restrictions.

- Try Odd and Even Ended Nim, Balloon Ride Revisited, and No Twin Nim.

Odd and Even Ended Nim

1–6

How

- Place 17 toothpicks (or other markers) in a row. Play with a partner.

- Take turns picking up 1, 2, or 3 markers.

- Play until all of the markers are gone. At the end, the person who ends up with an odd number of markers wins.

- This game can be played cooperatively. After you play several times, discuss strategies with your family for the best way to play the game. Family members can talk through their moves and share their thinking as they play.

- One way to play cooperatively is to start with 18 markers. Take turns picking up 1, 2, or 3 markers. At the end of the game, both players win if they have an odd number of markers.

- Try creating another Nim game where both players can win.

This is about understanding odd and even numbers and developing deductive reasoning skills.

MATERIALS
17 toothpicks or markers per two players

Up the Down Staircase

1-6

MATERIALS
cubes
pencil
crayons
paper
grid paper (optional), pages 178–179

MATH CONNECTION
Square numbers and square roots can be difficult concepts to understand when taught as rules to memorize. Here we build models using simple numbers and patterns to help us determine outcomes for larger numbers. We are beginning to build a foundation for looking at algebraic expressions in the later grades.

Remember the old lady who lived in the shoe who had so many children she didn't know what to do? Well, she's learned a few things. To keep order in her home she hired a contractor to build two staircases. This way her many children can go up and down the stairs at the same time without blocking each other's way.

How

- Build the following model and keep adding to the staircase until there are at least 5 steps up and 5 steps down the other side.

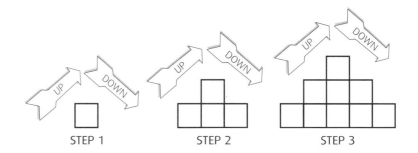

STEP 1   STEP 2   STEP 3

- How many cubes are there in a 5-step staircase? Describe how the number of cubes increases every time you add a new step.

- Organize the information to show how the number of cubes increases each time you add a step to the staircases.

STEPS	CUBES
1	2
2	4
3	9
4	?
.	.
.	.
.	.
10	?

- How many cubes would the contractor need to build a 10-step staircase? 20 steps? 50 steps?

This is about exploring square numbers by organizing information and creating patterns.

Here's More

- Younger children may try the 5-step staircase and stop there. If your child is interested and curious about the higher numbers, keep on building.

- How many cubes would you need for a 100-step staircase?

- If you only had 96 cubes, how high would the staircase be?

- Compare the outside stairs in the illustration above with what you've built. How are they the same? Different?

Baseball Cards

2-6

This is about
using algebraic thinking and logic to understand rates of change.

MATERIALS
one set of cards per game, pages 51–52
beans or markers

MATH CONNECTION
Baseball Cards helps children organize information, notice relationships between numbers, and think logically. These important skills are needed as algebra becomes more formal.

Ben keeps his baseball cards in sets. The following four sets will help you play with numbers and words. Each set has four cards with questions and clues. You and three family members or friends can use the clues to help Ben answer the questions on his baseball cards.

How

- Copy and cut out the baseball card sets on pages 51–52. Keep each set of four cards together.

- Take a set and give each person a different card.

- Read each card aloud, beginning with a question.

- Work cooperatively to answer the question(s). Discuss each of the clues and how they relate to each other. Make sure everyone gets the opportunity to participate.

- How do you figure out the answer to the question(s)?

- Now do the other sets. What do you discover as you read the clues and answer the questions?

SET 1
There are 24 Mariners cards in Ben's collection.

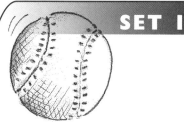
SET 1
Ben has twice as many Mariners cards as he does Astros cards.

SET 1
Ben has three times as many Astros as he does Giants cards.

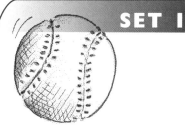
SET 1
How many cards does Ben have altogether in this set?

SET 2
Ben has 16 cards in all, Yankees and Red Sox only.

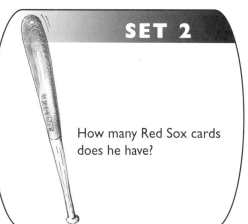
SET 2
How many Red Sox cards does he have?

SET 2
He has 4 more Red Sox cards than Yankees cards.

SET 2
How many Yankees cards does he have?

BASEBALL CARDS

BASEBALL CARDS

SET 3

Ben has twice as many Cubs cards as Padres cards.

SET 3

There are fourteen cards in all.

SET 3

He has half as many A's cards as Padres cards.

SET 3

How many of each card does he have?

SET 4

There are 21 Rangers cards in Ben's collection.

SET 4

How many rookie cards does he have?

SET 4

He has two times as many veteran cards as rookies.

SET 4
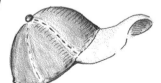
How many veteran cards does he have?

Baseball Cards

Here's More

Ben is 10 years old and lives in California. Once a year, he visits his granddad who lives in Boston. When he visits his granddad, he gets to choose three baseball cards from his granddad's collection. This tradition started when Ben was three years old.

The first time Ben visited, he picked three very old cards. The following year, he selected three new rookie cards.

- Figure out how many cards Ben collected after two years.

- How many cards will Ben have if he visits his granddad every year until he is 19 years old? If he keeps collecting at this rate, how many cards will Ben have when he is 25?

- Make a picture or graph that shows his baseball card collection growing.

BEN'S AGE	VISITS	TOTAL CARDS
3	1	3
4	2	6
5	3	9
6	?	?

For more collection activities, see *FAMILY MATH for Young Children: Comparing.*

Do you have a collection?

How do you choose what will go in it?

How do you organize your collection?

Make a display of your collection.

Paths and Ponds

2-6

MATERIALS
50-60 cubes or construction paper squares of two colors
crayons
paper
pencils
grid paper, 1" or 2 cm, pages 178–179

REAL-WORLD CONNECTION
The Moors in Spain loved water. Their garden environment reflected this in the many ponds and pools in their homes and gardens. The Alhambra in Granada, Spain is a wonderful example of how water was used to create serene environments that also had practical applications. For example, the Moors' waterwheel technology was adapted for use as ancient air conditioning.

The sight and sound of murmuring water is considered a critical element in designing gardens today. Let's explore the designs for a series of ponds. The following are ideas for building square and rectangular ponds.

How
Square Ponds

- Let's begin with cubes (or 1" paper squares) that represent 1-foot-square tiles. Use them to figure out how many tiles you need to make a border or path around the edge of a pond. Try using one color for the pond and a different color for the border. Don't forget the corners.

This is about connecting patterns and functions through the investigation of area and perimeter.

- Make some models for ponds that are 1´ by 1´, 2´ by 2´, 3´ by 3´, and 4´ by 4´.

1' BY 1' POND

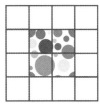
2' BY 2' POND

- Figure out how many square tiles you will need for the border of each pond.

- To record your work, use the grid paper and color in the squares. Use one color for the pond and a different color for the border.

- Make a table. See if you can figure out a pattern that will tell you how many tiles you need for the border of any size square pond. Talk about how to use the table to figure out the border for your ponds.

LENGTH OF POND SIDE	POND SQUARE FEET	# OF BORDER TILES
1	1	8
2	4	12
3	?	?
.	.	.
.	.	.
.	.	.

- Before you move on to the rectangular ponds, ask your child what might happen next. Does your child see any patterns?

Paths and Ponds

How
Rectangular Ponds

- Ask your child to create some rectangular ponds in the following sizes: 1´ by 2´, 2´ by 2´, 3´ by 2´, 4´ by 2´... Add a border for each pond using your paper squares or cubes.

- See if your child confidently moves the tiles into the new shapes. If not, let your child keep exploring square ponds until she is ready to move on.

- Make a table to record your results similar to the one for square ponds. Show the length (let the changing number be the length), the width, the area, and the border for each pond.

- Make another table for the 1´ by 3´, 2´ by 3´, 3´ by 3´ family of ponds. Work out the pattern for the number of tiles in the borders for these ponds. What patterns does your child see?

- Next try the 1´ by 4´, 2´ by 4´, 3´ by 4´ family.

Here's More
- Ask your child what would happen if you had 24 tiles for the border. Using all 24 tiles, what size ponds could you enclose? Don't forget the corners.

- Create a table for these ponds.

Patios and Paths
2-6

The concepts from this activity are relevant to making a real path or patio using tiles.

You may wish to do Paths and Ponds on page 54 before doing this activity.

How
- You have 16 square tiles to make a patio or path. Using this number of tiles, what patio and path shapes can you create? Keep track of your designs by recording their lengths and widths. How many were you able to make?

- Pick out a number of square tiles between 19 and 101 and use them to figure out what rectangular-shaped patios or paths you can make. For example, what could you build with 36 tiles? Make a table to show your results. Remember that a square is a special type of rectangle.

- Try another number, and compare your results with the others.

- Which type of numbers gives you the most patios or paths? Which numbers don't give you so many patios or paths? Can you explain why?

Here's More
- Use triangular tiles made from square tiles cut on the diagonal to design patios and paths. Try tessellating with the triangles. (See sidebar on this subject.)

- Make a design with squares and triangles by using different colors of paper.

- Another tiling project might be to make a tabletop from a certain number of leftover tiles.

This is about
connecting patterns and functions through the investigation of area and perimeter.

MATERIALS
50-60 cubes or construction paper squares of two colors
paper
pencil
grid paper, 1" or 2 cm, pages 178–179

REAL-WORLD CONNECTION
The Moors designed intricate and complex tessellations using geometric shapes in their tile work throughout the Alhambra in Spain. The work of M. C. Escher is another excellent example of tessellation in art. However, he used birds, fish, and other living things as his subjects.

To make a tessellation means to cover a flat surface (plane) with shapes to make a repeating pattern that does not overlap or leave spaces uncovered. You can use pieces that are exactly the same size and shape. You can also try mixing shapes like hexagons and triangles or squares and triangles.

Grampa's Coins

3–6

MATERIALS
real dimes, nickels, pennies, and quarters or play money

MATH CONNECTION
The ability to translate word phrases into algebraic expressions is crucial, not only in algebra but in almost all higher mathematics and science courses. This skill is also indispensable in economics, biology, geology, physics, and engineering.

Danielle and her brother Jordan have fun playing games with their grandfather. He has a habit of jiggling the change in his pockets when he walks. Grampa made up a game. If the children can guess the number of coins he has in his pocket, they can split the money. To help his grandchildren, he offers them some clues.

This is about
using algebraic language to develop number sense, proportional reasoning, and logical thinking.

How

- Work together to help Danielle and Jordan figure out the amount and number of coins in Grampa's pockets.

- Each pocket contains at least one of each coin: penny, nickel, and dime. There may be more than one combination of coins that will work for each pocket description.

- If your child gets stuck on one problem, leave it and move on to another one. Go back later to the problem that was difficult.

- Talk about how you solved each problem and describe the strategies you used.

1. This pocket contains:
   Half as many nickels as pennies
   4 dimes
   16 coins in all

2. This pocket contains:
   Twice as many nickels as dimes
   8 nickels
   85¢ in all

3. This pocket contains:
   An equal number of nickels and pennies
   The number of dimes is a square number.[1]
   $1.08 in all

4. This pocket contains:
   Two more pennies than nickels
   The value of nickels is four times the value of the pennies.
   $1.00 in all

[1] A square number is the result (product) of a number multiplied by itself. For example, 25 is a square number, a product of 5 x 5.

GRAMPA'S COINS

FAMILY MATH II: ACHIEVING SUCCESS IN MATHEMATICS

Grandpa's Coins

Each of the following pockets contains at least one of each coin: penny, nickel, dime, and quarter.

5. This pocket contains:
   The same number of dimes as nickels
   One-third of the coins are dimes.
   Five is a factor[2] of the value of each coin group.

6. This pocket contains:
   Half as many nickels as quarters
   The value of dimes and pennies together is half the value of quarters.
   10 coins in all

7. This pocket contains:
   A nickel for every two dimes (or a 2-to-1 ratio of dimes to nickels)
   The value of quarters is equal to the value of dimes and nickels.
   10 coins in all

8. This pocket contains:
   Three more nickels than pennies
   The value of nickels equals half the value of quarters and dimes together.
   The total number of quarters and dimes is half the number of nickels.

9. This pocket contains:
   One more nickel than pennies
   Two fewer pennies than quarters
   The total number of coins is a prime number.[3]

Here's More

- Make up your own problems. Hint: start with some coins and use them to help make up the clues.

- See Number Line Rectangles on pages 115–119 of the original *FAMILY MATH* book for more exploration of square numbers.

---

[2] A factor is a number that divides evenly into another number (with no remainder).

[3] A prime number is an integer bigger than 1 which has no factors besides 1 and itself. For example, 7 is a prime number because 7 and 1 are the only two whole numbers (factors) that will give you a result of 7 when multiplied.

"And women in turn...may encourage girls' desire for relationship and for knowledge and teach girls that they may say what they know and not be left all alone."

—Carol Gilligan, *Joining the Resistance Mother Daughter Revolution,* page 137

Royal Family Puzzle

4-6

MATERIALS
4 Aces, 4 Kings, 4 Queens, and 4 Jacks from a deck of playing cards

REAL-WORLD CONNECTION
Latin Squares have many real-life uses. In agriculture, a Latin Square arrangement can be used to determine the effects of different growing conditions. These conditions can include various types of soil, use of fertilizer, amount of light and water on the growth of several kinds of seeds.

A Latin Square arrangement can also serve as the basis of a plan that allows the testing of four different brands of tires on one car. The tires would be rotated weekly to observe the effects of wear in each of the four-wheel positions of the car.

Latin Squares have some interesting characteristics. The elements can be rearranged within the square in a variety of ways. For example, shifting rows or columns will keep the construction of a Latin Square intact.

A Latin Square is a collection of elements (numbers, letters, colors, etc.) arranged in a square. Any number of elements can be used as long as each appears only once in each row and column.

Ask your child to arrange the letters A and B in a 2 by 2 matrix so that each letter appears only once in every column and only once in every row.

Your child's matrix might look like this:

Now using the letters A, B, and C, ask you child to fill in the blank spaces on the 3 by 3 matrix.

A	B	C
B	C	A

If your child is confident, you are ready to move on to The Royal Family Puzzle. If your child is unsure, go back and work with the 2 by 2 or 3 by 3 matrix again or until your child feels confident.

How

- The object of this investigation is to place the cards in a 4 by 4 matrix (4 rows and 4 columns), so that each row and each column contains exactly one Ace, one King, one Queen, and one Jack. The cards can be of any suit.

- Work with a friend to make a matrix (layout) that follows the rules. When you are done, you will have what is called a Latin Square. Make a record of your solution.

- If you were to draw a line vertically down the middle and another horizontally across the middle, what cards would be in the four quarters of the arrangement?

- What cards are in the 2 diagonals?

- What other patterns do you notice?

- Create the matrix (arrangement) in another way. What cards are in the quarters and in the diagonals now? Compare the patterns.

- How many different ways can you make the 4 by 4 matrix? How do you know each one is different?

Here's More
- There are many types of Latin Squares, and the four-by-four square can provide many hours of problem solving enjoyment. A good investigation for older children would be to find all the different ways to construct a four-by-four Latin Square and to identify and analyze distinctive patterns.

- Create a new matrix considering both the denomination (A, K, Q, J) and the suit (♥, ♣, ♦, ♠) of each card. Each row and each column must contain an Ace, a King, a Queen, and a Jack and only one of each suit. Example of a possible row: A♥, K♣, Q♦, J♠.

- Now try a new Latin Square where not only must each row and column contain one of each Ace, King, Queen, and Jack and one of each suit, but also both diagonals must meet the same condition. How can the strategies you used in the previous arrangements help you?

- For younger children, start with two denominations, for example two Kings (♥, ♣), and two Queens (♥, ♣), to make a 2 by 2 matrix. Then make a 3 x 3 matrix using Kings (♥, ♣, ♦), Queens (♥, ♣, ♦), and Jacks (♥, ♣, ♦).

INTRODUCTION

Number Sense

> One, two, three, four, five,
> I caught a fish alive,
> Six, seven, eight, nine, ten,
> I threw it back again!
>
> *I Caught a Fish Alive,* a children's counting song

Remember the counting songs you sang as a child? Which counting songs have you taught your children to sing? Counting fingers and toes is often the first experience children have with numbers. However, what do young children really understand about numbers? While a three-year-old may be able to hold up three fingers and say, "I am this many," chances are the child doesn't understand what that really means.

This section is about numbers and making sense of them. The activities are designed to help children develop an understanding of how numbers work and interact with each other.

For many of us, developing number sense meant mastering the four basic functions of addition, subtraction, multiplication, and division. We memorized facts, often without understanding. To truly develop number sense, it is important to know why we add, subtract, multiply, or divide and what each means mathematically.

Developing number sense may include singing songs about number sequences. It also includes understanding the conservation of quantity—that is, that four items remain four items whether they are presented close together or spread out over a larger area.

In this chapter, you will find engaging challenges that will help your family explore numbers. Figuring out the way numbers work together is quite powerful. When children understand the idea of "threeness," they can make connections to other areas of learning that relate to threes. Triangles, triplets, tricycles, and trios become connected with meaning.

Having a strong foundation in number sense will help children as they move on to more complicated mathematics such as algebra and geometry. In FAMILY MATH, it is this deeper understanding and relevance that makes learning mathematics worthwhile.

"Because young children develop a disposition for mathematics from their early experiences, opportunities for learning should be positive and supportive. Children must learn to trust their own abilities to make sense of mathematics."

—*Principles and Standards for School Mathematics,* page 74

This is Six

MATERIALS
pencils
crayons
stickers (optional)

MATH CONNECTION
When children learn about numbers, they are often asked to see them as parts of a sequence or in isolation of other numbers. It is important for children to see that *one* number can be the result of various combinations of other numbers.

How

- Think of all the ways you can to express the number six (6). You can draw six of anything, but only once. For example, if you draw six stars, don't draw six bananas in another square. If you do, you will show six in the same way only with different objects. Try using a variety of numbers, symbols, pictures, words, or anything else. Remember, once you have shown a combination in any form (pictures or symbols) do not create the same combination again.

SHOW DIFFERENT WAYS TO EXPRESS THE NUMBER 6	
Example: 6 stars, separate sets	Example: 2 dogs, 4 cats (6 animals)
Example: 2 (separate) sets of 3	

- If you have more ways to represent six, draw them on a chart.

This is about exploring relationships between numbers and working with addition, subtraction, multiplication, and division.

Here's More

- Try this with different numbers from 1–9. What did you discover?

- Make a book titled *Our Book of Six* (or any other number you choose).

- Start a collection to express other numbers using pictures, stickers, or words.

Clean Your Plate!

K–3

MATERIALS
playing cards, all suits, #'s 1–5
 (Aces are worth 1)
a paper plate
30 beans, bottle caps, or wooden cubes
 per two people

MATH CONNECTION
This is a terrific way for younger children to practice the concepts of subtraction or addition. As children develop their computation skills, they will stop using the objects and do the math mentally.

A reminder: basic skills are an important part of mathematics and mastering them need not be tedious and boring.

How

Subtraction:

- Two people play using one plate. Ask your child to shuffle the cards and stack them face down.

- Decide who goes first.

- Place thirty beans on the plate and take turns drawing a card.

- When it is your turn, take off as many beans as the number on your card indicates.

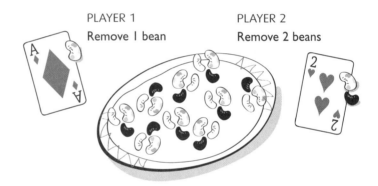

- To end the game, you must draw the exact number that "cleans the plate." If you run out of cards, just reshuffle them. Remember that you can only draw one card per turn.

- When the plate is empty, count your beans.

- If you like you can play a noncompetitive game and see who ends up with an odd or even number of beans. Or, the person with the most beans wins the game. You can also change the rules so the player who takes the least number of beans wins. Decide before you begin the game!

Addition:
- Two people play using one plate. Ask your child to shuffle the cards and stack them face down.

- The object of this game is to be the person who puts the last bean *on* the plate.

This is about counting, subtracting, and adding.

- Decide who goes first.

- Place 30 beans next to the plate and take turns drawing a card.

- When it is your turn, add the same number of beans to your plate as the number on your card.

- Play until one of you gets the exact number that "fills the plate." Remember that you can only draw one card per turn.

Here's More
- For younger children, begin with 10 beans and use the cards 1–3. Remember that Aces are worth 1.

Beaded Braids

INVESTIGATING PATTERNS & RATIOS

MATERIALS
Simone's and Brianna's braid boards, pages 72 and 73
40 hair beads or tag paper beads, page 181

MATH CONNECTION
Mathematics is about patterns. In the early grades, children observe, describe, copy, and create patterns. As children move through the grades, they investigate the relationship between elements in patterns.

REAL-WORLD CONNECTION
Patterns are designs or sequences that repeat in a predictable order. A wallpaper border might have a pink shell, a tan shell, and a gray shell, which appear in that order over and over. There are repeating patterns in music. For example the "one, two, three, one, two, three" pattern in a waltz is known as three-quarter time. A canon in music refers to the fact that the same melody is played over and over in counterpoints, inversions, and at different tempos. Engineers check to make sure that patterns in creating products are consistent. This assures that, for example, the gasket that comes off the assembly line at the beginning of the day looks just like the one that comes off at the end of the day.

Some patterns are simple like the pink, tan, and gray shells; and some are complex like the canons in music. Here we investigate bead patterns for Simone and Brianna's braids.

Simone loves to wear beads in her braids. She has four braids. She always wears more heart-shaped beads than round ones, and she likes lots of beads.

Brianna likes both heart and round beads, but she prefers only a few beads on each of her two braids.

Whenever their mom braids the girls' hair, she has to sort the beads out ahead of time to be sure that there are enough beads to satisfy both Simone and Brianna.

How

- Cut out 20 round and 20 heart-shaped beads from page 181, or you can use beads you have at home.

- Start with Simone's braids. Use the board on page 72 to create a pattern. Work with a partner to divide some beads between her four braids. Make sure the pattern is the same for each braid, and that she has more heart beads than round ones. Don't forget, Simone likes lots of beads on each braid.

- Arrange the beads so that each braid will have the same number of beads in the same pattern on each of Simone's four braids. Don't forget to leave some beads for Brianna.

- Now sort the remaining beads for Brianna's two braids. For this you will need the board on page 73. Make sure that the pattern for each braid is the same.

This is about designing patterns to explore ratio through addition, multiplication, and division.

- How many arrangements can you find to please both girls? Illustrate your bead patterns so that you can keep track of the ones you have created.

- What can you say about the beads on one of Simone's braids?

- Sometimes there may be some leftover beads. What does this mean about the pattern?

Here's More
- What would happen if Simone had five braids or six braids?

- If Simone had 100 braids, how many beads would she need?

- How can you express the pattern in terms of ratio for one of Simone's braids? For example, for one of her braids, how many heart beads are there for each round bead?

- What would the ratio be if you used 15 round and 24 heart-shaped beads on three braids?

Just For Fun
- Select your favorite colors and color the tag paper beads in sets of 20. For example, you might color 20 red hearts and 20 yellow round beads. Or, color 20 blue heart beads and 20 green round beads.

- Color the game boards.

- Practice making braids with yarn. It is fun and helps develop fine motor skills.

To show the ratio between the round and heart beads, we write Number of round beads: Number of heart beads. For example, the ratio between 8 round beads and 12 heart beads is written 8:12.

SIMONE'S BRAIDS

BRIANNA'S BRAIDS

BEADED BRAIDS

FAMILY MATH II: ACHIEVING SUCCESS IN MATHEMATICS 73

FAMILY MATH on the Go

K-6

There are opportunities for teaching and learning everywhere we go. Here's a list of mathematical things your family can try when you are traveling:

- Estimate how many minutes or seconds it will take to arrive at school in the morning. Measuring time, how do walking, driving, or riding the bus compare?

- Estimate how long it will take to get somewhere, and give your child a watch to keep track of the time. By watching the time and "checking" on the accuracy of everyone's estimation, your child will develop a sense of time and will practice telling time.

- Talk about the turns you are making as you drive. Hearing you say "Now we are turning left," or "At the next corner we'll turn right," helps your child establish a sense of direction. Older children can try to determine whether you are turning east or west, or going north or south. Look for landmarks that can be used as visual references.

- Ask your child to count the traffic signs. Are there any signs in the shape of an octagon? Which ones are shaped like a rhombus or a triangle? You can also look for signs by color. Which sign do you see most often?

- Is there an extra long block on the way to the store or market? Have your children compare the lengths of two blocks. For example: Which is larger? Which is shorter? How can they check their guess?

- Play "I Spy" when traveling long distances. For example, if your child sees a car with a number that is the same as her age, she can say "I spy with my little eye, a car with the number 5 on it!" Keep track by making check marks or putting stickers on a paper. At the end of the trip count how many cars you saw that fit this rule. You can also add characteristics, such as favorite color: "I spy with my little eye, a blue car with the number 5 on it."

- Have older children choose any number as a factor. Every time they see their number, they multiply it. For example, if you chose 6 and you spot a number 6 on a license plate, you multiply the two sixes and get 36. The next time you see 6, you multiply 36 x 6 and your new answer is 216. Keep this going until you reach your destination. You might want to keep a calculator in the car. As a treat, someone will open the car door for the person who scores the highest number.

- Sing songs such as "18 Wheels on a Big Rig," "The Ladybugs' Picnic," or other number songs you know.

Wonder about things aloud. Show delight in the pursuit of understanding and problem solving. When your children see you as an inquisitive, curious person, they acquire similar behaviors.

Box Math

ADDITION & SUBTRACTION

MATERIALS
Box Math Game Boards + and –, pages 78–81
number and operation cards, page 174
scratch paper
pencils

MATH CONNECTION
This activity reinforces the patterns of addition and subtraction. The more we understand how numbers can be put together and taken apart, the easier and more accurate arithmetic computations become. This process is sometimes called *composing* and *decomposing* numbers.

How

- In this activity, you get the answers first. Your challenge is to place the number cards and operation cards (+ or –) in the correct spaces.

- Copy on card stock the number and operation cards (+ or –) displayed on page 174, and cut them out. Later you can store them in envelopes or small plastic bags.

- Figure out where the number cards go and which operation card you will use to get each answer. Place the number cards in the boxes and one of the operation cards (+ or –) in the circles. To solve each problem, use each number only once.

- Start with the Box Math A Game Board. Arrange the number cards (3, 5, and 7) with one operation card (+ or –) to get 42. Discuss how you worked out the answer.

ANSWER: 35 + 7 = 42

- Now try 28. What do the answers tell you about what numbers go in which boxes?

- Finish the Box Math A Game Board.

- What do you know about the relationships between the answers, the operation cards (+ or –), and the set of numbers?

- Now try Box Math B.

- When your child feels confident with Box Math A and B, try Box Math C and D.

This is about
applying logical thinking and mental math to explore number patterns and place value using addition and subtraction.

Here's More

- Choose some different numbers and make new problems to share with your family.

- Try the Box Math Multiplication and Division activity on page 94.

You will find *Box Math A* numbers (3, 5, 7) and operation cards (+ or −) on page 174.

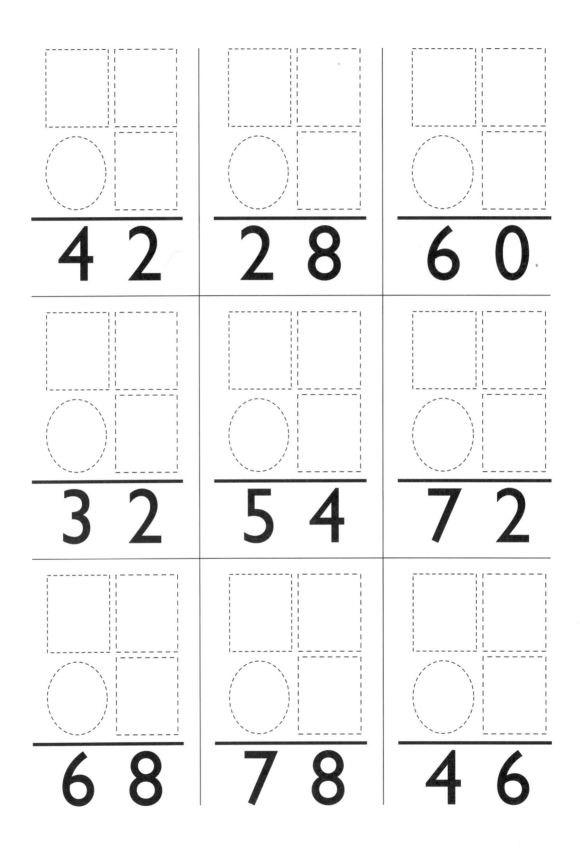

You will find *Box Math B* numbers *(4, 6, 8)* and operation cards *(+ or −)* on page 174.

You will find *Box Math* C numbers (3, 5, 7, 9) and operation cards (+ or –) on page 174.

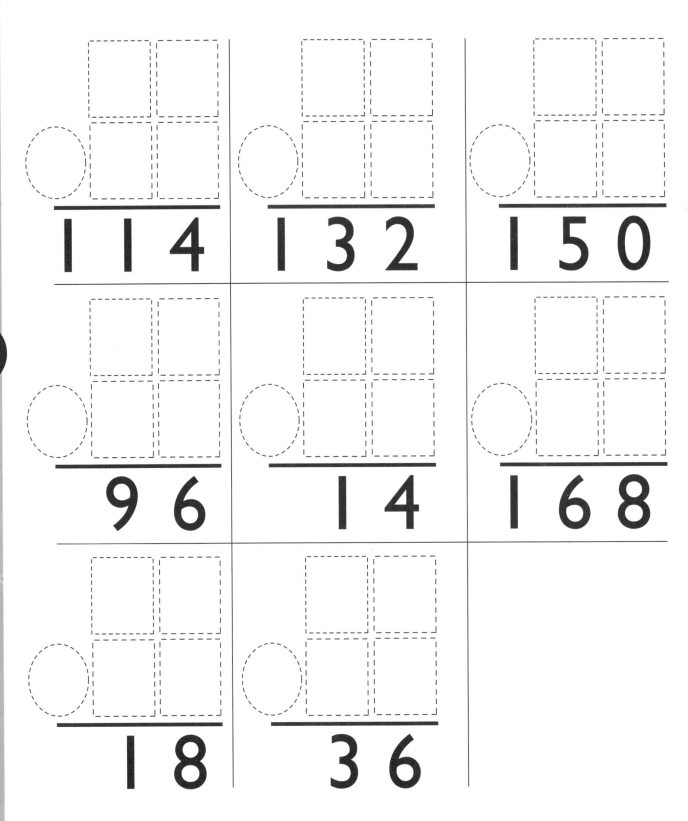

You will find *Box Math D* numbers (2, 4, 6, 8) and operation cards (+ or –) on page 174.

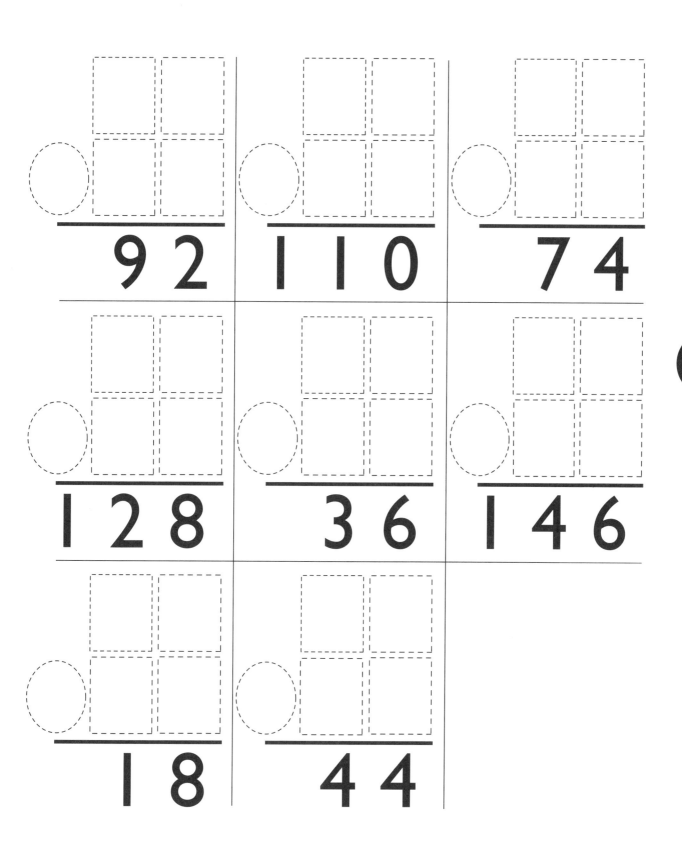

Consecutive Sums
2-6

MATERIALS
pencil
paper
calculator (optional but helpful)

REAL-WORLD CONNECTION
Looking at sequences in events is related to understanding electricity, computer programming, robotics (a form of computer programming), and even knitting machines!

Before You Begin

For this activity we will use whole numbers that go in order. Whole numbers that go in order without skipping a number are called *consecutive numbers*.
   For example:
   1, 2, 3, 4 are consecutive numbers.
   So are
   12, 13, 14, 15.

Numbers that are *not* consecutive are 1, 3, 4, and 5 because the 2 is missing.

Which number is missing in this nonconsecutive sequence?
   30, 31, 32, 34, 35

Which of these are sums of consecutive numbers?
   1 + 2 + 3 = 6
   2 + 2 + 4 = 8
   4 + 5 = 9
   1 + 4 + 5 = 10
In the example above, we notice 4 + 5 = 9. Is there another way to write 9 as the sum of consecutive numbers?

How

- Using only whole numbers (no fractions or decimals please), show all the possible ways to find each number from 1 to 25 (or 1 to 10 for younger children) as the sum of consecutive whole numbers.

   Here are some examples for adding two consecutive numbers:
      1 + 2 = 3
      2 + 3 = 5
      3 + 4 = 7

- How much does the sum grow each time?

- What is the difference between two sums?

- Record the sums you find. Be sure to include the addends you used. What patterns do you notice?

This is about
adding consecutive whole numbers to explore the relationship between addends (the numbers we add) and their sums.

Here's More

- What happens to the patterns when you add three consecutive numbers? What about four consecutive numbers?

- Share your findings with one another. How are your solutions different?

- How were the patterns for the sums similar or different? For example, how did the pattern for adding two consecutive numbers differ from the pattern for adding three consecutive numbers?

Bean Boxes
3–6

MATERIALS
scratch paper for drawing boxes
45 beans

MATH CONNECTION
Have you ever tried to divide something equally? Bean Boxes is a rich investigation about creating groups with the same number of items in each one. At first, it looks like we are just dividing. However, the rules in this activity limit us and we find ourselves using logic to solve each new problem. Looking for patterns in numbers is another way to explore number relationships in multiplication.

How

- Two or more family members can do this activity together. Please allow a lot of time to explore and discuss as you go through each step.

- The goal is to rearrange the beans so that each box has an equal number. Keep a record of each step you try.

- Draw 3 boxes and use them to distribute 12 beans so that there is a different number of beans in each box, as shown in the following example. You may want to start with this arrangement.

- Arrange the beans so that each box has an equal number of beans, following these rules:

    1. Anytime you move beans from one box to a new box, you must put into the new box the exact number of beans that are already there.

    2. Although you may move beans from any box, all of the beans moved at each turn must come from a single box. In the example above, you could move two beans into the left box from either of the others, or move four beans into the center box from the right-hand box. Could you move the two beans in the left-hand box anywhere?

- How many moves did it take you to equalize the boxes? What is the smallest number of moves you can make to equalize the boxes?

- Try starting with a different arrangement of 12 beans. In fact, you might want to figure out how many possible starting positions there are.

This is about developing an understanding of factors and multiples by using logic.

- How many arrangements can you create? Do all arrangements work to equalize the boxes? What patterns do you see?

- When you know all you want to know about 12 beans, you are ready to try a different number.

- Choose a number between 12 and 45 and see if your number works with this model. Why do you think you can use some numbers and not others? Try to find a pattern or general rule for what will or won't work.

- How can you predict what the least number of moves might be to equalize the number of beans in each box?

Remember, you are trying to make the boxes hold an equal number of beans in the fewest number of moves.

A factor is a number that divides evenly into another number (with no remainder).

A multiple is the product of a whole number and any other whole numbers. A multiple of 4 is 8; 8 is a multiple of 2.

Doubling Bean Boxes

3–6

MATERIALS
scratch paper for drawing boxes
45 beans

You may want to do Bean Boxes before trying this activity.

How

- Work with your child, taking turns and keeping a record of each step.

- Begin with some number of beans, divided *unequally* into two boxes. For example, start with 10 beans, divided into one box of 3 beans and one box of 7 beans.

- Double the size of the smaller group by taking 3 beans from the larger group, leaving new numbers of 4 and 6 beans in each box.

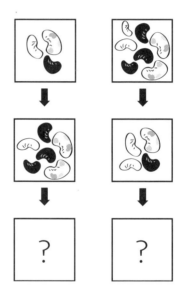

- Using the new numbers of beans in the boxes, repeat the process. In other words, double the size of the new smaller group by taking beans from the larger group. What are your new numbers?

- Continue the process. What is happening?

- But that's not all! Try this again but this time begin with 5 and 6 beans in each box.

- Keep track of each step. Compare what happens when you start with each new set of numbers.

86 FAMILY MATH II: ACHIEVING SUCCESS IN MATHEMATICS   NUMBER SENSE

- Try other pairs of numbers such as 3 and 5, 7 and 1, and 2 and 14.

- Try larger pairs of numbers such as 2 and 24, 16 and 4, and so on.

More Questions
- How many moves are made in each case before the smaller group becomes the larger?

- Look for patterns with all the possible combinations for the numbers 7 and 20. What patterns do you find?

- Do number sets that are close together and those far apart give different results?

- Sometimes you reach a point where no moves can be made. How many moves did it take to get there?

- Which numbers are unusual?

- Do the factors of a number matter? How do you know?

This is about
developing an understanding of factors and multiples by using logic.

Loopy Lou I

MATERIALS
adding machine paper
pencils
calculator (optional)
scratch paper

MATH CONNECTION
How do math rules happen? How can we know when something we do in math will always have the same outcome? Mathematicians find particular examples of how numbers work persuasive, but they are always looking for a proof or a chain of logic that points to a new rule. A persistent soul is always looking for a proof.

Lou loves patterns and looks for them everywhere, especially in numbers. Recently he discovered a pattern that always leads him to 4, 2, and 1 as the final three numbers. He hasn't had time to try all the numbers he'd like to. Please help Lou figure out if his pattern works for any number.

How

- Work together; one person will record, and one will compute.

- Choose a number from 10 to 99. Follow the rules below. Use rule A or B on the number you choose, depending on whether it is odd or even.
  A. If your number is even, divide by two (÷ by 2).
  B. If your number is odd, multiply it by three (x by 3) and add one (+1).

This is about investigating number patterns through multiplication, division, and addition.

- Record your results and continue the process until you see a pattern or until you get to number one (1) as your final answer.

  For example: Here is a start if you chose the number 17.

	17 (odd number)
17 x 3 + 1 =	52 (even number)
52 ÷ 2 =	26
26 ÷ 2 =	13
13 x 3 + 1 =	40
40 ÷ 2 =	20

  Continue until you reach one (1) as your answer.

- Look at your record sheet. What kinds of patterns do you notice?

- At what point were you able to predict what numbers would come next?

- Why does this Loopy Lou always get you back to the same three numbers (4, 2, 1)?

Here's More
- Try this with different numbers. What happens when you multiply and divide using different numbers?

Loopy Lou II

3–6

MATERIALS
spinner, page 177
scratch paper
calculators (optional)

Lou's dad was so impressed with the number patterns Lou found that he wanted to show them to Lou's mom. However, she was busy working on a new number pattern. "Look at this," she said, "I wonder if Lou has discovered this pattern."

Try the following to see what Lou's mom discovered.

How

- Make a spinner from page 177.

- Each person selects two numbers from 0 to 9 by spinning twice.

- Add the two numbers, and write down the digit in the one's place.

For example:
If you spin a 3 and 9, you add them and get 12. Keep the 2, which is in the one's place. So now you have 3, 9, and 2.

- Add the last two numbers, (the 9 and the 2) and you get 11. Keep the 1, which is in the one's place. Now you have 3, 9, 2, and 1.

- Add the last two digits (the 2 and the 1), and that gives you 3. So now you have 3, 9, 2, 1, and 3.

- Keep adding the last two numbers until you get to 3 and 9 again (which are the original two numbers).

- How many numbers in your loop before the pattern repeats itself? Why might this be occurring?

This is about
practicing addition, understanding place value, and discovering number patterns.

Here's More
- What else can you use besides spinners to determine your starting numbers?

- Try a new pair of starting numbers.

Extension:
- Here's another model you can try. Start with 4 and 2. Let the next number be the remainder (R) or number left over when the sum of the previous two numbers is divided by 5.

Example:
4
2
          4 + 2 = 6;   6 ÷ 5 = 1 R1
1
          1 + 2 = 3;   3 ÷ 5 = 0 R3
3
          3 + 1 = 4;   4 ÷ 5 = 0 R4
4
          4 + 3 = 7;   7 ÷ 5 = 1 R2
2

Note, the pattern starts over after four numbers. A mathematician might say that this pattern has a period of 4.

Try two numbers, (0–9) of your own and see what happens.

Two Coin Problem

3-6

MATERIALS
pencils and scratch paper
two colors of markers to represent coins
hundreds chart, page 182
calculators (optional)

REAL-WORLD CONNECTION
Has your family ever been in another country? If so, how did you figure out how to count and spend money that is different from what you normally use? If you have never visited another country, choose a country that you want to visit in the future. How does that country's money differ from ours? In Two Coin Problem, we visit an imaginary country and learn how to make change using their money system.

How

- Imagine that your family recently returned from a country that used only 5¢ and 13¢ coins. Using only these types of coins, what priced items can you buy with exact change at the local shops?

- Work with your family to figure out if you can buy something at every price from 1¢ to $1.00? Discuss how to divide up the task.

- Use the hundreds chart on page 182 to record the possible prices. As you work, look for patterns in the mathematics. Discuss other ways to record your findings.

- What prices did you find that could not be made exactly with 5¢ and 13¢ coins? Share strategies with one another for arriving at certain numbers.

- Are there more or fewer possible prices than you expected?

- Are there some prices that you can get in more than one way? What are they?

- Do your findings reflect Frobenius' theory described in the note below?

Our FAMILY MATH colleague, Steve Jordan, from the University of Illinois in Chicago, pointed out that this activity is an example of a more general mathematics problem from number theory. Ferdinand Georg Frobenius (1849–1917) found that when you have a situation like this problem, there is a number beyond which you can always make exact change. He called that number the *conductor*. To find the conductor, you multiply one less than the first number by one less than the second. In this activity the conductor would be 4 x 12 or 48 (one less than five is four and one less than 13 is 12). Frobenius also proved that the number of gaps, or numbers you can't find, is equal to half of the conductor. In this case, the number of gaps would be 48 ÷2 or 24.

This is about
developing number sense and logical thinking, exploring multiples, and organizing information.

Here's More
- If the shopkeeper can only give you change using 5¢ and 13¢ coins, what other-priced items can you buy? For example: You can buy something for 2¢. You pay three 5¢ coins (15¢) and the shopkeeper gives you back a 13¢ coin for change. Or if you pay the shopkeeper two 13¢ coins for a 21¢ item, she can give you 5¢ change.

- For a more complex problem, try this with three numbers that do not have a common divisor. For example: the only coins you can use are 5¢, 13¢, and 17¢.

Box Math

MULTIPLICATION & DIVISION

MATERIALS
Box Math Game Boards, pages 96–99
number and operation cards,
  pages 175–176
scratch paper
pencils

How

- In this activity, you get the answers first. Your challenge is to place the number cards and operation cards (x or ÷) in the correct spaces.

- Copy on card stock the number and operation cards (x or ÷) displayed on pages 175–176, and cut them out. Later you can store them in envelopes or small plastic bags.

- Figure out where the number cards go and which operation card you will use to get each answer. Place the number cards in the boxes and one of the operation cards (x or ÷) in the circle. To solve each problem, use each number only once.

- Start with the Box Math E Game Board, and arrange the number cards and operation card to get 92. Discuss how you worked out the answer.

$$\square\square\bigcirc\square = 92$$

ANSWER: 46 x 2 = 92

- Now try 23. What do the answers tell you about what numbers go in which boxes?

- Working together, look at different ways to solve the rest of the problems in Box Math E.

- How did you use your estimation skills?

- When you see answers with decimals, what information does this give you that helps solve the problem?

- Now try Box Math F, G, and H.

This is about applying mental math and logical thinking as we investigate number patterns, place value, and decimals.

Here's More
- Choose some different numbers and make new problems to share with your family.

BOX MATH E

MULTIPLICATION & DIVISION

You will find *Box Math E* numbers (2, 4, 6,) and operation cards (× or ÷) on page 175.

☐ ○ ☐ ☐ = 252

☐ ○ ☐ ☐ = 7

☐ ○ ☐ ☐ = 248

☐ ○ ☐ ☐ = 15.5

☐ ○ ☐ ☐ = 104

☐ ○ ☐ ☐ = 6.5

☐ ○ ☐ ☐ = 92

☐ ○ ☐ ☐ = 23

☐ ○ ☐ ☐ = 128

☐ ○ ☐ ☐ = 32

☐ ○ ☐ ☐ = 144

☐ ○ ☐ ☐ = 4

BOX MATH

MULTIPLICATION & DIVISION

☐ ◯ ☐ ☐ = 558
☐ ◯ ☐ ☐ = 15.5
☐ ◯ ☐ ☐ = 207
☐ ◯ ☐ ☐ = 23
☐ ◯ ☐ ☐ = 288
☐ ◯ ☐ ☐ = 32

☐ ◯ ☐ ☐ = 324
☐ ◯ ☐ ☐ = 4
☐ ◯ ☐ ☐ = 567
☐ ◯ ☐ ☐ = 7
☐ ◯ ☐ ☐ = 234
☐ ◯ ☐ ☐ = 6.5

You will find *Box Math F* numbers (3, 6, 9,) and operation cards (x or ÷) on page 175.

BOX MATH

MULTIPLICATION & DIVISION

You will find *Box Math G* numbers (2, 4, 8) and operation cards (× or ÷) on page 176.

☐ ◯ ☐ ☐ = 328
☐ ◯ ☐ ☐ = 20.5
☐ ◯ ☐ ☐ = 168
☐ ◯ ☐ ☐ = 42
☐ ◯ ☐ ☐ = 24
☐ ◯ ☐ ☐ = 96

☐ ◯ ☐ ☐ = 192
☐ ◯ ☐ ☐ = 3
☐ ◯ ☐ ☐ = 5.25
☐ ◯ ☐ ☐ = 336
☐ ◯ ☐ ☐ = 7
☐ ◯ ☐ ☐ = 112

☐ ○ ☐ ☐ = 208
☐ ○ ☐ ☐ = 13
☐ ○ ☐ ☐ = 90
☐ ○ ☐ ☐ = 22.5
☐ ○ ☐ ☐ = 108
☐ ○ ☐ ☐ = 27

☐ ○ ☐ ☐ = 120
☐ ○ ☐ ☐ = 4.8
☐ ○ ☐ ☐ = 210
☐ ○ ☐ ☐ = 8.4
☐ ○ ☐ ☐ = 100
☐ ○ ☐ ☐ = 6.25

You will find *Box Math H* numbers (2, 4, 5) and operation cards (× or ÷) on page 176.

BOX MATH

H MULTIPLICATION & DIVISION

Family Garden

4–6

MATERIALS
two pieces of 8 $\frac{1}{2}$" by 11" paper of different colors
scissors
Family Garden Record Sheet, page 103

At the Komai annual reunion, Jelani's family gathers for spring planting. Each person brings a favorite vegetable or fruit to plant. They have decided to divide the garden area into sections. The oldest person gets $\frac{1}{2}$ of the whole garden to plant. The next oldest gets $\frac{1}{2}$ of what is left and so on. The family keeps dividing the garden area until everyone has planted something.

Grandmother Kiku, the oldest, chooses strawberries to plant this year. The whole family pitches in to help her, because everyone loves her homemade strawberry ice cream. The next oldest in the family, Uncle Danny, will take half the remaining area to plant corn. He plans to donate his corn to the community food bank.

What happens as they continue to divide the garden area? How would you describe what is happening to another person?

How

- Work together and select two pieces of $8\frac{1}{2}$" by 11" paper of different colors to make your garden. One person will cut and one will record, or you can alternate roles.

- Before you begin, label 1 piece of paper "1 whole." This will be the template for the whole garden and is where you will place your sections of fruits and vegetables.

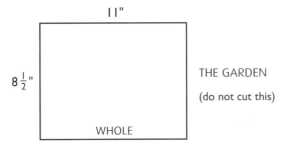

- Take the second piece of paper and fold it in half. Cut along the fold line. Draw your favorite fruit or vegetable on one of the pieces. Put the other piece aside for now.

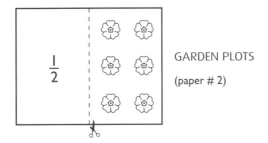

- Together figure out the fractional value of the first piece that you cut, and label the fraction on the back of this piece. Now you are ready to place this piece on the garden plot (template).

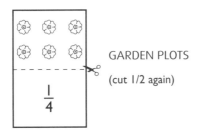

MATH CONNECTION

Making a physical model helps us understand the relative values of fractions. Children need to understand and compare the different values of fractions such as $\frac{1}{8}$ and $\frac{1}{4}$. This is especially true when children move to decimal fractions like 0.125 and 0.25. As the denominator (the bottom number of the fraction) increases and the numerator (top number) stays the same, the quantity represented decreases. For example $\frac{1}{45}$ is much smaller than $\frac{1}{4}$.

Family Garden

There are several ways to find the value of a particular fraction. One strategy is to ask how many times this piece will fit on the whole. Your answer becomes the denominator. For example, the largest fractional piece you cut fits on the whole template two times. So the denominator is two, and we write the fraction for the first piece as $\frac{1}{2}$.

- Take the piece you set aside, and repeat the procedure above, drawing a different fruit or vegetable for this garden section and labeling its fractional value. When you are finished, place this piece in your garden.

- Continue the process until the pieces become too small to handle. Remember to label your garden sections on both sides.

- How many different types of fruits and vegetables were planted? What would happen if you kept dividing the garden?

- Using the record sheet, make a list of the fractional pieces that make up your garden, arranging them from largest to smallest.

- What patterns do you notice? What is the relationship between the denominators and the size of the different pieces?

Note: Save your garden pieces and template. Use them for the Family Garden Game on page 104.

FAMILY GARDEN
RECORD SHEET

Record what you and your partner do in the following ways:

GARDEN SECTIONS	WHAT HAPPENED	FRACTIONAL VALUE
Template does not get cut	My original whole garden plot	1 whole
	Example: I took the second piece of paper and cut it into two equal pieces. I labeled one piece with my favorite fruit and its fractional value on the other side.	$\frac{1}{2}$

FAMILY GARDEN

Family Garden Game

4-6

MATERIALS
paper template and fractional garden pieces from the Family Garden activity
1 spinner, page 105
1 paper clip
1 pen or pencil

In order to play this game, you first need to complete the Family Garden activity on page 100.

How

- Copy and make the spinner on page 105 for this activity.

- Work with 2-6 players. Each player will need a set of fractional parts and a whole template from the Family Garden activity.

- Take turns using the spinner.

- Take the fractional piece selected by the spinner, and place it on your whole template.

- If you have already used the fractional piece selected, you can spin again 1 time. If you get the same fraction again, then you skip your turn.

- Each person must take five turns. The person with the smallest area uncovered has planted the most garden.

- Play again. Did you cover more area this time?

Here's More

- Start with your template covered with all your fractional pieces.

- Take turns using the spinner.

- Remove the fraction piece indicated by the spinner.

- If you have already used the fractional piece selected, you can spin 1 more time. If you get the same fraction again, then you skip your turn.

- The person with the largest area uncovered on their whole template after 5 turns wins the game.

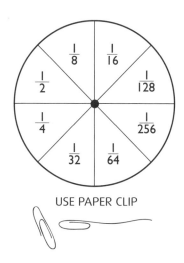

USE PAPER CLIP

This is about using probability and comparing fractions.

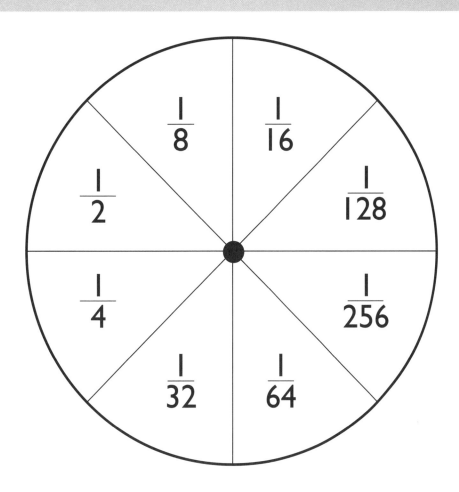

HOW TO MAKE YOUR SPINNER:
1. Make a copy of the above spinner.
2. Unfold a paper clip so that one edge is pointed straight as shown below.

3. Place the looped end of the paper clip on the center of the spinner.
4. Hold the paper clip in place with a pen or pencil.
5. Spin the paper clip by flicking the pointed end with your finger.

FAMILY GARDEN GAME

INTRODUCTION

Geometry

What do you think of when you remember your experiences with geometry in school? When some of us think of geometry, we remember high school geometry with its many theorems, rules, and proofs. Others might remember platonic solids and formulas for figuring out volumes.

Geometry is everywhere. It's in snowflakes, buildings, trees, and even in space. As a matter of fact, geometry is the study of shapes and space.

All children enjoy moving shapes around; they like stacking shapes until they fall over. When young children play with triangles, squares, circles, and other shapes, they experience the differences and similarities of these shapes firsthand. These early experiences help children develop the complex skills of understanding connections between shapes and visualizing patterns.

In this chapter, we experiment with a basic building element in geometry—the line. In Spaghetti Geometry, we visualize lines that go beyond the page to better understand how lines are continuous and how they behave. In Sidewalk Math, we use sidewalk chalk to create line segments. We use practical applications such as matting favorite pictures to learn about area and perimeter. Other activities invite us to build small structures by using reasoning and logic skills to explore surface area and volume.

As you try out these activities with your children, remember that the process is just as important as the final results. Woven through the activities are questions designed to generate discussions and extend thinking. When you engage your child in these mathematical conversations, your child is:

- taking risks in a safe environment with you as a guide
- using mistakes as springboards to new learning
- sharpening critical thinking skills
- learning how to talk about mathematical events
- thinking aloud about the understanding at hand
- building a bank of resources for becoming a self learner

These experiences provide important groundwork that will help children when they encounter the coordinate system and basic ideas of symmetry and proportion. Please refer to the original *FAMILY MATH* book for excellent activities in both these areas.

Having positive experiences with geometry early on will help children cultivate curiosity about the math we see all around us.

"Communicating about mathematical ideas is a way for students to articulate, clarify, organize, and consolidate their thinking."

—Principles and Standards for School Mathematics, page 128

INTRODUCTION

Sidewalk Math Using Sidewalk Chalk

108 FAMILY MATH II: ACHIEVING SUCCESS IN MATHEMATICS   GEOMETRY

This section includes adaptations of things to do outdoors on driveways or sidewalks. Using chalk (the kind you can wash away) we can create larger playing spaces so that children can move about as they practice arithmetic skills.

Small children enjoy doing this with other kids, but mostly they like having parents play with them. Let your child do as much of the drawing as possible. Resist the urge to correct their writing or the uniformity of their grids.

These activities are fun to try out with your child's playgroup. If more than three or four children are involved, be sure to draw several "game boards" so they won't have to wait too long in between turns.

Body Trace can be messy! Even if you don't want to be the one who lies down on the sidewalk, it is an engaging way to explore estimation, to observe the repeating patterns, and to try your hand at being artistic. For young children, the world is an extension of themselves. Their understanding of how things function begins with and extends from the self. Try this activity; be mathematical and creative!

Line Segments and Intersections explores line segments and intersections in a nonstructured, open-ended way. Everyone's line segments will be different, yet the questions asked about them create some common understanding. At the same time, everyone can make predictions about possible intersections, lengths, and outcomes.

Me and My Shadow introduces children to making connections between time, the movement of the earth, and shadows. The activity requires a time lapse between measurements. This connection to astronomy is a simple way to show children how mathematics exists in our everyday life.

It All Adds Up! is an inviting way to practice basic addition skills, sharpen mental math skills, and develop spatial reasoning. You can subtract instead of add, or you can multiply.

Value Line allows children to think about questions as they develop their language skills in a mathematical context.

In mathematics, the more answers you get, the more questions you generate. Encouraging children to talk about what they like or what they think about something also develops their confidence to ask questions and to express themselves.

SIDEWALK MATH
Line Segments & Intersections
K–3

MATERIALS
sidewalk chalk; a different color for each player
2–8 wooden cubes or small pebbles

MATH CONNECTION
This activity is a way to learn about line segments. Line segments have end points that separate one segment from another. When a child studies geometry in later grades, lines and intersections become part of polygrams, grids, and other more complex designs.

How

- Each person holds a chalk of a different color.

- Have your partner mark where you are standing so that when you move, the next person can stand on the same spot.

- Hold two wooden cubes or two small pebbles in your outstretched hand and let them drop on the sidewalk.

- With your chalk draw a straight line from one wooden cube or pebble to the other. Mark each end of the line with a dot to indicate the end point of your line segment. Remove cubes or pebbles.

- Have your partner stand in the exact same spot you did and drop two cubes and connect them with a line. Try making your line as straight as you can. Don't forget to mark the end points with a dot.

This is about creating and exploring line segments.

- Now stand back and compare your lines. How are they the same? How are they different?

- Are the lines as long as you are tall? How can you tell? Do your line segments intersect (do they cross over each other)?

- Keep doing this until you have created three lines each. Use a different color chalk than your partner uses so you can keep track of your own lines.

- How many intersections (points where line segments cross each other) do you have now?

Here's More
- Try dropping 3 or 4 cubes at a time. Talk about what shapes you might create, and then connect the points.

- Try dropping the cubes from different heights. How far apart do they land? Try again. What is different this time?

- What might happen if the two of you stand back-to-back and drop your cubes or pebbles at the same time?

- What would happen if you drop marbles instead of blocks?

- Look around you. Where do lines and line segments occur in nature?

SIDEWALK MATH

Me and My Shadow
K-3

MATERIALS
sidewalk chalk
kitchen timer or watch

REAL-WORLD CONNECTION
Observing events in the natural world and connecting them to mathematics provides a way to see the relationships between the patterns, cycles, and passing of time.

Although we have an agreed-upon standard for measuring time, various cultures have employed many ways to mark the passing of time in specific and consistent units. Examples of this are Chinese water bowls, French sand clocks, and calendars from other cultures and civilizations.

HOUR GLASS

SUNDIAL

How

- This activity is best done on a sunny day. Any number of people can do this activity together. Draw an X on the sidewalk.

- Stand in the sunlight and make a shadow. Make sure your shadow is on the sidewalk or driveway.

- Ask your partner to trace your feet where you are standing.

- Draw a straight line from your feet to the end of your shadow. Write the time of day at the end point.

This is about measuring and comparing the length of shadows over time.

- Do this again in half an hour. Don't forget to stand in the exact same spot and face the same way you did earlier. Write the new time of day at the end point. How much did your shadow change?

- If you can, measure your shadow once every half hour for two hours. For accuracy, set a timer or use the timer on the kitchen stove if it has one. For example, measure your shadow at 2:00, 2:30, 3:00, and 3:30 p.m.

- What do you suppose is happening?

- Is your shadow long? Short? Describe your shadow.

- Measure the shadows of other objects. What happens? What happens when we measure the shadow of a tree, for example?

SIDEWALK MATH

Body Trace

This is about
practicing estimation while learning about number and measurement.

MATERIALS
yarn
sidewalk chalk
ruler or yardstick
yourself

MATH CONNECTION
Estimation is an everyday skill. How much time will you need to get ready for school or work? How much time will it take to get there? Which grocery store line will get you through faster? How much will the groceries cost? How many miles is it from here to the next town? The more you estimate, the better you are at it.

How

- Estimate how much yarn it will take to outline your body.

- Lie down on the sidewalk, and have another family member trace around your body with chalk.

- Lay the yarn around the perimeter of the body shape. Was your estimate close to the actual measure?

- Now guess how much yarn it will take to outline your partner's body and repeat the process.

- Whose yarn is longer? Shorter?

- How close were your estimates?

- Just for fun, fill in your features and clothes.

SIDEWALK MATH

It All Adds Up!
K–6

This is about practicing mental math using addition or multiplication.

How

- Draw a 3-square-by-3-square grid on the sidewalk. Write the numbers 1–9 in squares like this:

FOR PRIMARY GRADES

0	3	5
5	1	2
2	4	1

FOR INTERMEDIATE GRADES

1	4	7
8	2	6
5	9	3

- Take about ten big steps back from the grid and draw a starting line.

   TOSS MARKERS

   FROM THIS LINE

- Toss two markers one at a time onto the grid you created. Add the two numbers on which the markers land. It's OK if both markers land on the same number, just add that number to itself or multiply times 2. Record your sum after each round. Remove your markers so that the next player can play. Repeat until each of you takes five turns.

- Add up your five scores and see how large a number you get. Now add everyone's scores together. How large is your combined number?

- Play again. This time try for the highest individual score. How did you change your strategy in the second game?

Here's More

- Make up a new rule to play the game. For example, you could make the rule that only one marker may land in a square. If your marker lands on an occupied square, it equals zero.

- Try multiplying instead of adding.

- For older children, let the first marker you toss be the numerator and the second marker be the denominator of a fraction. Then add the resulting five fractions for your total score.

MATERIALS
sidewalk chalk or masking tape (for indoors)
bean bags or other hopscotch markers
paper and pencil (optional)

It is important to use numbers and operations (adding, subtracting, multiplying, or dividing) that your child can do successfully. Add new or more challenging skills when your child is ready.

SIDEWALK MATH

Value Line
K-6

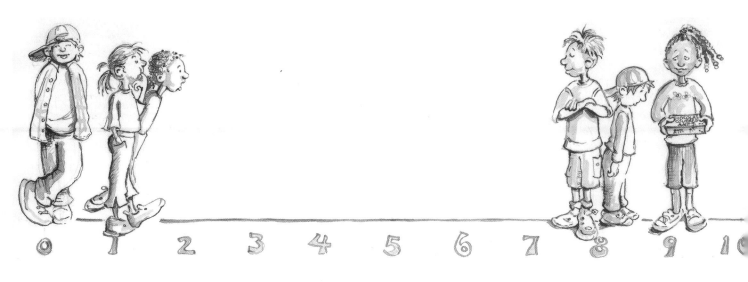

MATERIALS
chalk
yourself and a couple of friends or family members

REAL-WORLD CONNECTION
Likert scales, graphing information, and making value judgments are aspects of collecting information and assessment. They are used for a variety of purposes such as marketing or providing products to specific communities or groups.

How

In this activity, we will use a value line to get children thinking about their opinions on particular issues.

- Draw a straight line about 10 feet long on a sidewalk or driveway. Write numbers on it from 0 to 10, like this:

Figure out how to make these numbers evenly spaced without a yardstick.

- Let 0 stands for NO WAY! and let 10 stand for ABSOLUTELY!

- Think of a question that requires an answer on a scale between 0 and 10. Ask each person (including yourself) to stand on the number that best represents his or her answer.

This is about understanding sequence on a number line.

Here are some questions you can ask. Or, you can make up you own.
- I would eat chocolate-covered ants.
- I would ski on a snowy day.
- I would sing on a stage for a large (100 people) audience.
- I would like to plant a garden.
- I would eat ice cream until I was ready to explode.

- After everyone has settled on where they want to stand, look around. Compare how many people are on each of the numbers. Discuss your differences of opinion.

- Here are some more statements, but in this case 0 stands for NEVER! and 10 stands for A LOT!

  - I go to the library.
  - I use the internet for homework.
  - I watch TV on the weekends.
  - I dance.
  - I recycle cans, glass, and paper.
  - I speak another language besides English at home.
  - Make up some more questions of your own and HAVE FUN!

Name Game

MATERIALS
pencils
paper

REAL-WORLD CONNECTION
Venn Diagrams are a basic thinking tool for understanding sets within sets (set theory). The logic used in Venn Diagrams is very important in science, advanced mathematics, and the design of computer software.

Venn Diagrams are drawings that show relationships among sets of objects. They are named after an Englishman, John Venn. He lived until 1923 and made these symbolic drawings popular. Although they are usually drawn with circles, other shapes such as squares and triangles can be used. Venn Diagrams are a helpful way of sorting things into categories.

Some letters in the alphabet are written with only straight lines; others are only curved lines. Some letters have both straight and curved lines.

How

- Working together, print your names on a scratch sheet of paper.

- Compare your name with other family members' names. Does one name have more curved letters than the other?

- How are your names similar? How are they different?

- Write all the letters with only straight lines in the circle at the right.

- Write all the letters with only curved lines in the circle on the left.

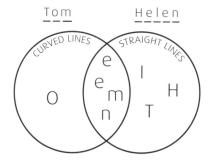

- Which are the letters with both curved and straight lines? Where should they go in the Venn Diagram? What else can we say about the letters that share both characteristics?

- How else might we organize the information to show the same results?

- Is there another kind of diagram or graph that would be easier to read or understand? Create it!

- Try this with your best friends' names!

This is about developing a logical way of sorting letters with curved and straight lines by using a Venn Diagram.

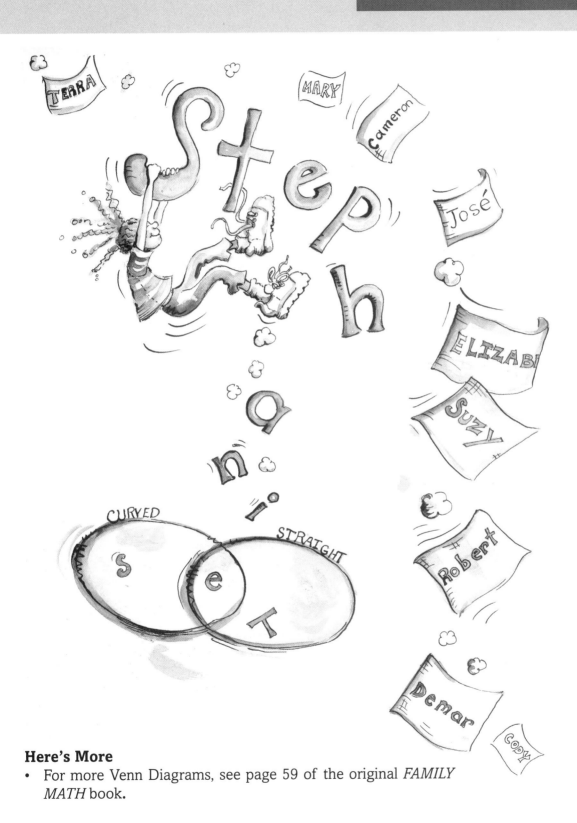

Here's More
- For more Venn Diagrams, see page 59 of the original *FAMILY MATH* book.

Dot to Dot

MATERIALS
crayons
color pencils or felt pens
paper
rulers (optional)

REAL-WORLD CONNECTION
When one line crosses another line, it is said to intersect that line. We use lines and intersections to create maps and to survey, as well as to mark points in space.

The famous artist Paul Klee used lines of varying color with interesting intersections in his art. This activity invites us to be creative as we explore lines and intersections.

How

- Work in pairs. Decide who goes first. One person chooses a crayon or color pencil and draws two dots anywhere on a piece of paper.

- Using a different color, the second person draws a straight line to connect the two dots and then places two more dots on the paper. The person who went first connects these dots using a different color. These two new dots must result in a line that crosses over the first line. This will create the first intersection.

- Alternate drawing dots and connecting them with straight lines until each of you has drawn three to four lines. Remember that all lines must have at least one intersection. Don't forget to make the dots one color and lines a different color.

- How many points of intersection do you have? To make counting them easier, you can circle each intersection as you count it.

- What are some true statements about your design?

Here's More
- Try connecting three dots instead of two. What happens to the lines? How does this affect the intersecting points?

- Try to draw a straight line that crosses over as many other lines as possible. How many intersections does it have?

- If you especially like any of your drawings, display them prominently. See the Matting Pictures activity on page 136 to enhance your artworks.

Line Symmetry

K–6

MATERIALS:
piece of string about 12" long
symmetry cards, pages 124–125
scratch paper
color crayons

Rotational Symmetry

Imagine that a circle has been imposed over a square. Find the center of the imaginary circle. That point is also the center of the square. If the square can be rotated around that point and fits into its outline one or more times before it reaches the starting point, then we say that it has rotational symmetry. We divide the number of times it fits into its outline into 360° and that gives us the degrees of rotation. So we say that a square has 90° of rotation because it can fit into its outline 4 times (360/4) if it is rotated around a central point. You can check rotational symmetry of other shapes in the same way.

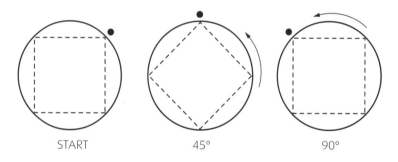

START   45°   90°

If you can fold a figure in half so that the two sections match exactly, the figure has a line of symmetry.

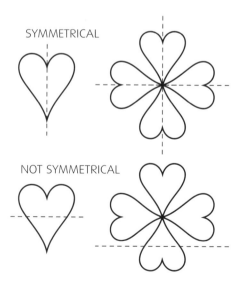

SYMMETRICAL

NOT SYMMETRICAL

> **This is about**
> investigating symmetry and using visual and spatial skills.

How

- Work with a partner. Pick a symmetry card and use your string (stretch it out) to determine how many lines of symmetry each shape has. Move the string around or move it from side to side.

- As you move your string around, have your partner verify that a line (or lines) of symmetry exists.

- How do you know when a shape does or does not have symmetry? How can you prove it?

- Name some things in your home that have lines of symmetry.

- Print your name. Which letters in your name have line symmetry?

- Try finding the lines of symmetry in other letters of the alphabet.

- What is unique about the circle and its symmetry?

Here's More

- Create some of your own shapes that have line symmetry.

- What percent of the letters of the alphabet have lines of symmetry?

- For more symmetry activities, see the original *FAMILY MATH* book.

LINE SYMMETRY — DESIGN CARDS

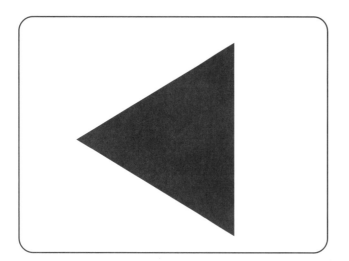
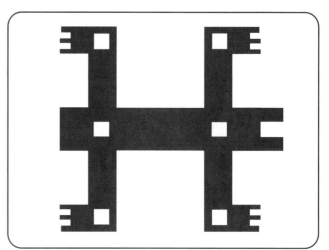

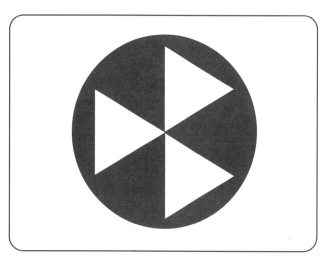

LINE SYMMETRY

Many Faces of a Cube I

K-6

MATERIALS
one set of eight small cubes prepared as described below
sticky dots

REAL-WORLD CONNECTION
Spatial visualization skills are important in everyday life. Reading and sketching maps, and giving and following directions are examples of spatial thinking in action. Have you ever struggled with the instructions for putting toys or furniture together? Spatial reasoning helps us make sense of those instructions.

Preparation
- Prepare a set of eight cubes. Fit the small cubes together to form a big cube as illustrated.

- Keeping the big cube intact, place a sticky dot, a triangle, or any distinguishing mark on each exposed side (or face) of the smaller cubes. All exposed faces including those on the bottom should be marked.

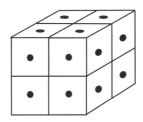

- If you do not have 8 cubes available, here's a pattern for making them out of paper.

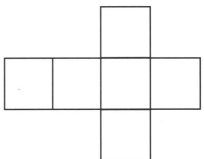

This is about
practicing spatial reasoning and logical thinking using surface area.

How

- After you have labeled all the faces, mix up the cubes. Rebuild the big cube by putting the small cubes back together. Remember that all the marked faces must be exposed.

- How are the big cube and small cubes the same? How are they different?

- What is the relationship between a small cube and the big cube?

Here's More
- Try Many Faces of a Cube II.

Many Faces of a Cube II

K-6

MATERIALS
one set of 27 cubes prepared as indicated below
sticky dots

REAL-WORLD CONNECTION
When pilots land an airplane, when surgeons perform a heart transplant, when engineers design a bridge, and when astronauts repair a space station, they all make use of strong spatial reasoning skills.

You may want to try Many Faces of a Cube I before trying this activity.

Preparation

- Fit 27 cubes together to form a larger cube.

- Keeping the larger cube intact, place a sticky dot, triangle, or any distinguishing mark on each exposed side (or face) of the smaller cubes. All exposed faces including the bottom should be marked.

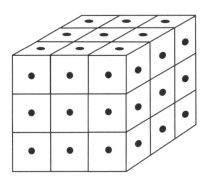

This is about
building big cubes with smaller ones and using the constructed cubes to explore surface area and volume.

- If you do not have 27 cubes available, here's a pattern for making them out of paper.

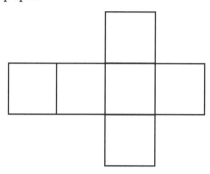

A surface area is the total area of the faces (including the bases) and curved surfaces of a solid figure.

Volume is the number of cube units of space a solid figure takes up.

How

- After you have labeled all the faces, mix up the cubes. Rebuild the larger cube by putting the small cubes back together. Remember that all the marked faces must be exposed.

- Predict how many cubes will have 0, 1, 2, and 3 sides marked. Compare your thinking with each other. Count the faces on the cubes and see how close your predictions were.

- Talk about the relationship between the increase in volume and the increase in surface area. What happens to surface area and volume as the size of the big cube increases or decreases? How can you show this?

- How many small cubes will be necessary for the next larger cube? What can you say about the number of marked sides on the smaller cubes?

- What patterns do you see as you build larger cubes?

Here's More

- It might be helpful to create a table to see any patterns.

SIDE LENGTH	# OF DOTS SHOWING LARGER CUBE	# OF CUBES IN LARGER CUBE VOLUME
1	6	1
2	24	8
3	54	27
4	?	?
5	?	?

Measurement & Estimation

2-6

MATERIALS
1 or 2 rulers, 30 cm or 12"
   meter and yard sticks
metric and standard measuring tapes
paper and pencils
objects to measure

MATH CONNECTION
Estimation is a very important skill in mathematics. It is particularly important in measurement. This activity provides an approach to making more accurate estimates when measuring length.

REAL-WORLD CONNECTION
Quite often you need to make a good estimate of a measurement when you don't have a ruler with you. This can happen when you want to know if a table on sale will fit next to your bed or if a picture frame is big enough for your new 11" by 17" poster. Many people use a part of their body such as a hand or foot to help estimate measurements in these situations.

How

- Ask your family if they know some measurements that they use to help them make estimates without a ruler.

- Discuss how you could use body measurements or "units," such as hand span or stride, to help you make estimates when you don't have a ruler.

- Make some measurements of your own body "units." Study the measurements to see which come out close to whole inches or centimeters. Make a note of the results you might want to use to help make estimates.

   Some examples of body units to include are:
   width and/or length of little finger
   hand span
   arm span
   foot length—with a particular pair of shoes or barefoot
   stride
   fist circumference
   floor to waist

This is about becoming more accurate in estimating, measuring, and organizing information.

- Can you think of other body units that would be helpful?

YOUR BODY UNIT	ACTUAL MEASUREMENT

- Measure the length of some objects using your body units. Then compare those measurements by using a ruler or tape measure.

OBJECT	BODY UNIT	ESTIMATE	ACTUAL MEASURE
FAMILY CAR			
BATHTUB			
YOUR BED			
FAMILY PET			
SKATEBOARD			

* Don't forget to make an estimate first.

Here's More
- Find out how things were measured before rulers.

- Take a survey of your friends and acquaintances. Ask them how they estimate measurements when they don't have a ruler or other measuring tool with them.

Spaghetti Geometry

3–6

MATERIALS
uncooked spaghetti or flat-sided toothpicks
chart, page 135

MATH CONNECTION
Doing geometry often requires the ability to visualize relationships of objects or lines in space.

REAL-WORLD CONNECTION
Lines are everywhere we look. They make art and architecture appealing. The geometric design in natural phenomena like snowflakes, trees, and flowers can be illustrated with lines.

A plane is a flat surface that extends infinitely in all directions.

Let's explore how many ways lines—that's lines, not line segments—can intersect in a plane (a flat surface). Lines extend outward in both directions, so potential intersection points must be imagined.

A whole line is infinitely long and can only be visualized. You can only model part of the line or see it in a drawing. Here we use pieces of long, thin, uncooked spaghetti or flat-sided toothpicks to investigate intersection points. Then you will make sketches to show what has been discovered.

A line is a straight path of connected points that continues without end in both directions. A line has length but not width. One way to draw a line is like this:

The arrows indicate that the line goes on.

Parallel lines are lines that are in the same plane and never cross because they are always the same distance apart.

Intersecting lines are lines that *do* cross. They may have one point in common (the point where they cross) or all points in common.

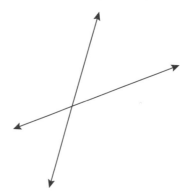

If they have *all* points in common, they become one line, that is to say they are the same line.

This is about understanding lines and intersections in geometry.

How

- Divide the tasks below among family members. One person can arrange the spaghetti lines on a flat surface, and the other person can draw the arrangement in the appropriate box on the chart from page 135.

- Begin with one piece of spaghetti. Draw a line in the box marked zero intersections to represent this single piece. As you can see, one line has no intersecting points, so let's move on to two lines.

- Arrange two pieces of spaghetti so they don't intersect. Draw this in the box marked zero intersections.

SPAGHETTI GEOMETRY   FAMILY MATH II: ACHIEVING SUCCESS IN MATHEMATICS   133

Spaghetti Geometry

- Rearrange the two pieces of spaghetti so one piece will cross over the other creating one point of intersection. Show this in the one intersection box. Can two pieces of spaghetti have two points of intersection? Why or why not?

- Keep adding spaghetti pieces and noting the numbers of possible intersecting points. Draw the new arrangements in the appropriate boxes. You may want to take turns so that each person gets to illustrate and arrange the spaghetti pieces.

- Look for patterns or relationships between the number of lines (spaghetti pieces) and the number of intersections as you build new arrangements.

- What other ways can you show this information?

For younger children, have them explore lines and intersections by providing materials to create line art using toothpicks, spaghetti, or other objects. Invite your child to tell you about the different lines in the finished artwork.

Spaghetti Geometry is based on an idea presented by Glenda Tinsley at a National Council of Teachers of Mathematics (NCTM) Canadian Conference.

SPAGHETTI GEOMETRY

# OF SPAGHETTIS (LINES)	ILLUSTRATIONS OF INTERSECTING POINTS			
1	0 INTERSECTION POINTS	1 INTERSECTION POINT		
2	0	1	2	3
3	0	1	2	3
4	0	1	2	3
	4	5	6	?
5	0	1	2	3
	4	5	?	?

Matting Pictures
4-6

MATERIALS

pencils, scissors, and rulers
construction paper
pictures, postcards, and photos
right triangles for measuring
   square corners
glue, scotch tape, or double-sided tape

REAL-WORLD CONNECTION

Picture framing and mat cutting call for a thorough understanding of fractions and symmetry. A good frame can enhance a picture dramatically. A good framer needs mathematics and a good eye for color and design.

Have you ever thought about framing a photo or picture? Usually there is a border or mat that surrounds the picture, sometimes two overlapping mats of different colors. How does the mat cutter know exactly where to cut to get the right-sized hole with the right look? In this activity you will learn how to cut mats that fit different sizes of frames.

Some common sizes of picture frames are 11" by 14", 8" by 10", and 5" by 7". The dimensions refer to the width and length of the frame. These dimensions are often referred to without mentioning the inches, as in an "eleven by fourteen frame."

With a little practice, it's not hard to work out the mathematics of cutting a mat. Mat board is very heavy and requires a utility knife or special cutter, but you can get good results making a mat using construction paper. Even if you don't have a frame, the finished product can look attractive.

Let's Practice Cutting Your First Mat

How

- For practice, we'll cut an opening with a one-and-a-half-inch ($1\frac{1}{2}$") border out of 8" by 10" paper. Can you predict the dimensions of the opening?

Mat cutters recommend making all measurements from the same corner after that corner has been carefully squared off.

- Measure and cut an 8" by 10" piece of paper. Be sure that the measurements are accurate and that the corners are square.

- "Measure twice and cut once" for your border, and discuss with your family how to make sure your measurements are accurate before cutting.

- Draw pencil lines to guide your cutting. In order to make sharp corners and preserve paper or mat board, start your first cut in the middle of a line and extend the cut a little past each corner. These slight overcuts will not show at a distance, and they prevent fuzzy corners.

- Check your accuracy by measuring the dimensions of the opening in several places for each direction.

Matting Pictures

Making a Mat with a Uniform Border

How

- The directions here are for a 4" by 6" picture or postcard to be put into an 8" by 10" frame.

- Pick a favorite 4" by 6" photo, postcard, or picture to mat. Make sure the corners are square.

- Choose a piece of construction paper that is a light neutral color or one that picks up a color in the picture.

- Measure carefully and cut the construction paper to 8" by 10". Square the corners carefully. Mark an X lightly in pencil in one corner.

- Calculate the opening for the picture in the mat. As you work, remember that the artwork needs to be bigger than the opening. Mat cutters allow the mat border to overlap the artwork by $\frac{1}{8}$" on all sides. One way to think of this is to think of the horizontal and vertical dimensions of the opening as each $\frac{1}{4}$" less than those of the picture—which means you subtract $\frac{1}{4}$" from each dimension. So, for a 4" by 6" picture, you will need a $3\frac{3}{4}$" by $5\frac{3}{4}$" opening.

BACK SIDE OF MAT. DOTTED LINE INDICATES MAT BORDER.

PICTURE OVERLAPS MAT BORDER

- Now work out the dimensions of the borders. Remember, you want the same dimensions for all of the borders—on each side and on the top and bottom.

- Talk about how to do this calculation. Some people focus on the opening when they do this calculation; others focus on the borders.

- Measure carefully and draw the lines for the opening, working from the corner pencil-marked X. Cut out the opening, overcutting at the corners as you did in the practice piece.

- Line up your artwork with the mat, and tape it in place. If you have a frame, put the matted artwork into the frame.

Making a Mat with a Double Uniform Border

- Some picture framers like to use two mats of different colors, one cut a little smaller than the other to create a double border effect. The mat with the smaller opening is normally a darker color.

The two mats are usually cut so that $\frac{1}{4}$" of the inside mat shows on all four sides of the picture.

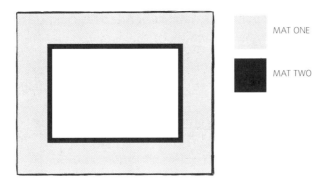

- Select a picture and two colors of construction paper for a double-border mat.

- Calculate what size opening you need for the picture, and cut that opening in the inside mat. *Remember* that the artwork needs to be bigger than the opening. Allow the inside mat border to overlap the artwork by $\frac{1}{8}$" on all sides.

- Then calculate the dimensions for the outside border so that the opening is $\frac{1}{4}$" larger than the inside mat on each of the four sides. This means the opening is $\frac{1}{2}$" larger in length and $\frac{1}{2}$" larger in width than the inside border. Cut the opening.

Matting Pictures

- Check to see if the borders line up OK. If they are off, you may need to cut a new mat. Don't be discouraged. This is tricky to do the first few times.

- Line up the two border mats and art, and glue in place. For best results with tape, use rolled up scotch tape or double-sided tape. Put your work in a frame if you have one.

Here's More
- Use different sizes of pictures and frames.

- Cut a mat with openings for two or three photos or post cards to go in the same frame.

- Some people like to have the bottom border of a mat be different from the top border. Pick a picture and make a mat for it with top and bottom borders that differ by at least $\frac{1}{2}$". Try the picture with a wider border on top and then on the bottom. Which do you like best?

For Younger Children

- Pick out a 4" by 6" picture or postcard. Make sure the corners are square.

- Pick out a piece of construction paper that looks good with your picture. Cut it to 8" by 10".

- Lay the picture on the construction paper so that it is "centered." Measure to see that the picture is the same distance from each side and the same distance from the top and bottom. Adjust accordingly. When it is centered, mark each corner spot on the construction paper lightly with a pencil. "Mount" the picture on the construction paper by gluing it in place or carefully placing several pieces of rolled up scotch or double-sided tape on the back of the picture and repositioning it. Check that the picture stays centered.

- If you have a frame, place the picture in the frame.

- If you like, decorate your mat using crayons, stamps, or stickers.

Organizing a FAMILY MATH Class Series

Good news! There are many ways to organize and lead a FAMILY MATH class. The class may be taught by parents, teachers, and other community members—the only requirement is a desire to share mathematics in an enthusiastic and nonthreatening way. We've included a number of suggestions gained from our own and others' experiences in starting the program, but you are the expert on what will work best in your own community.

WHO SHOULD COME?

Grade levels

An early decision involves what grade level(s) you want to recruit. Some people prefer offering a class for only one or two grade levels, since they think it's easier to present activities that will hold the interest of all the children. On the other hand, this is FAMILY MATH, and parents might want to bring an older or younger sibling. Many classes have been successful addressing a K–6 or 4–8 range. Groupings also work well: K–2, 3–4, and 5–6. Pick what suits you best, based on your interest and knowledge of the math topics and the needs of your community.

Parents and children

It's important to be very clear that your FAMILY MATH classes are for parents (or other interested adults) who may attend with or without a child, but that NO CHILD MAY ATTEND WITHOUT AN ADULT. If you do not insist on this rule, you may find yourself with a very successful tutorial session for a lot of students who want extra help or enrichment activities. This is not the purpose of FAMILY MATH.

WHEN IS THE BEST TIME?

Scheduling

Any time of the year is a wonderful time to present FAMILY MATH classes. However, there seems to be greater interest in attending FAMILY MATH classes during the fall and early spring months. In the fall, parents want to help their children get started in school on a positive note. In the spring, children who are having problems with math can be helped by attending a series of FAMILY MATH classes.

Naturally, you'll want to avoid school vacations, holidays, and other busy times. As to time of day, we have found it very difficult to maintain an afternoon class, but the early evening has nearly always been successful. Sometimes a Saturday morning class has worked. Before you set a time for your class, you may want to send home a recruitment notice with the children you are targeting, including a response portion for parents to indicate their preferred time.

FAMILY MATH
CLASS SERIES

LEARN

How to help your child with math at home.

What math your child will learn in school this year.

How to make math fun.

TAKE HOME

Materials, games and activities, career information.

MEET

Other families from your community.

People who use math in their jobs.

For more information, call:
_____

Please return the bottom of this sheet to:
_____
_____
_____
_____

- -

Yes, I'm interested in FAMILY MATH.

_____
NAME

_____
ADDRESS

_____ _____
TELEPHONE   BEST TIME TO CALL

_____
CHILD'S NAME

_____ _____
CHILD'S SCHOOL   CHILD'S GRADE

_____
CHILD'S TEACHER

Best time for me to attend a FAMILY MATH class:

Weekdays: ❏ (1:00–3:00 p.m.)   ❏ (3:00–5:00 p.m.)   ❏ (6:00–8:00 p.m.)
Saturdays: ❏ (10:00 a.m.–12:00 noon)

_____
FIRST CHOICE

_____
SECOND CHOICE

❏ I cannot attend a FAMILY MATH class series now, but please keep me on your mailing list.

Organizing a FAMILY MATH Class Series

WHERE TO HOLD THE SESSIONS?

A school classroom or library, church, community center, YMCA, YWCA, or science center are all good places for a FAMILY MATH class. Some things to look for:

- A facility that can be used for free, or for a small fee.
- Enough tables and chairs for everyone, with some for extra activities.
- A separate room for very young children and a babysitter.
- Ample parking.
- Safety for night classes.

Wherever you hold your sessions, make sure it is a place that is easy for parents and families to get to.

HOW DO I GET PEOPLE TO COME?

Recruiting

This can be the easiest or the hardest part of the series. If you know what a typical turnout is at your school for open houses, PTA meetings, and school plays, you may have some idea how difficult or easy it will be. If your community typically has high parent involvement, you may have more people interested than you can handle. If the parents in your community do not attend school functions but are regular churchgoers, your best strategy may be to recruit through churches. You may also include parents from other classes or schools.

Be aware that it is possible to over-recruit. Decide how many people you are willing to work with. If you think you might be overwhelmed with responses, start small, recruiting from parents of one or two classes.

Notices

A good procedure is to ask the teachers of the grade levels you want to attend to send home FAMILY MATH notices similar to the sample with their students. Phone calls make good reminders. You may want to emphasize in your notices that FAMILY MATH will provide parents with a chance to:

- learn about their children's math curriculum
- practice problem-solving skills with their children that will help them succeed with this curriculum
- encourage their children to persist in mathematics when it is no longer required

"When students compute with strategies they invent or choose because they are meaningful, their learning tends to be robust—they are able to remember and apply their knowledge."

—*Principles and Standards for School Mathematics*, page 86

Organizing a FAMILY MATH Class Series

Public announcements

Don't forget radio and newspaper announcements if you want to reach a large audience. Be sure to include a telephone number.

Sample Radio Spots

01 September 02
FOR IMMEDIATE RELEASE (class starting September 17)
Public Service Announcement—30 seconds

If you're the parent of a son or daughter in kindergarten through grade 6, you should know that one of the most important actions you can take as a parent is to help your children to enjoy math and encourage them to keep taking it through high school. Even if you don't feel like an expert in math, you can help your child.

Come to a free FAMILY MATH class Wednesday nights, starting September 17, at Rosa Parks Elementary School in Berkeley. For information, call 510.555.1823

01 September 02
FOR IMMEDIATE RELEASE (class starting September 17)
Public Service Announcements—10 seconds

Berkeley parents of elementary students! Help your kid succeed in math! Come to free FAMILY MATH classes held Wednesday nights at Rosa Parks Elementary School, Berkeley. For information, call 510. 555.1823.

Group presentations

One of the most effective ways to attract families to your class is to offer to make a presentation at a PTA or other parent group on "How to help your child with math at home." In twenty minutes you can explain the program, share suggestions for math activities that are easy to do at home, and do at least one math activity with the group. Collect names and phone numbers of interested parents. Bean Boxes is a good activity to do in a short presentation because it is interesting to adults as well as children and gets everyone involved. Be sure to bring copies of the activity you present, enough for everybody, with flyers to inform them about the class.

FINANCIAL SUPPORT

Finances

Make a list of your possible expenses to be sure you can cover your costs. Although it is possible to charge a "per family" fee, it should be kept as small as possible; and free is always preferable. Don't be afraid to ask for donations — from principals, businesses, philanthropic organizations, parent groups, your local grocery store or gas station. Most people find the idea of FAMILY MATH very exciting and are willing to contribute materials, food, or money.

Costs

Here are some of the expense items to consider:

Handouts and materials—probably about $5.00 per family for a six-week session.

Rent—try to get space free in a school, church, or community learning center.

Refreshments—you may be able to get them donated, or parents may be willing to take turns bringing them. Some school monies can be used to provide refreshments for all meetings where families are attending.

Your own time—you may be able to arrange a small salary through adult school or community college sources. Grants can cover stipends for parents, and the school Parent Teacher Organization can arrange for volunteers. The best way to fund class leaders' time is to include family learning in the school's overall budget.

SETTING UP

Preparation

See the Planning Check Sheet on page 156. Be sure you have all the details arranged, including enough handouts for everybody, all the materials for each activity, and a seating arrangement where you and the participants can see and hear each other.

Setup

Give yourself plenty of time to set up—be ready for the early birds. For the first session, you may need to arrive an hour or more before the official time. To help everyone get to know each other, provide name tags. Have juice or coffee available. If you want to keep attendance—and it's a good idea for record keeping—ask people to sign in on a Venn Diagram or Birthday Graph.

You can explain that a Venn Diagram is a way of sorting and classifying information and that these diagrams can be very simple or complex. You can also explain that graphs show lots of information in an organized manner.

You may also want to maintain a more formal attendance list, with addresses and phone numbers so you can maintain contact with the families.

OK, THEY'RE ALL HERE—NOW WHAT DO I DO?

Welcome

Be warm, be nurturing, be energetic! The atmosphere that you establish is the most important part of the class. A nonthreatening, comfortable environment gives parents and children the confidence to take some risks, try something new, and make mistakes. The activities themselves will generate discussion. Encouraging people to ask questions and talk about their thinking will help them feel free to speak out.

The most important messages you want to convey to parents are not necessarily put into words, but are conveyed through our own actions and beliefs—that mathematics has great beauty, that everyone can learn math, and that it can be a pleasure to do mathematics. When families can begin to concentrate on the fun of doing mathematics rather than getting the right answer, your class will be a success.

LET'S BEGIN

Openers

A typical FAMILY MATH class begins with an activity that can be set out with little or no explanation as an "opener" for people to do as they arrive. It's important to begin the class at the designated starting time, but you don't want parents to feel uncomfortable if they are late. You can begin a discussion about the opener about five or ten minutes after the official start of class.

Program description

At your first meeting, you'll want to explain the FAMILY MATH program, make introductions, and give an overview of your series of classes. For your first activity, pick something that is based on arithmetic skills, since this is the kind of math that most adults remember from elementary school. Later, when you introduce an activity from a less familiar math area, you can talk about the variety of math topics covered in school today and their interrelationships.

Stations

Math stations allow families to explore independently. These are activities arranged around the room with self-explanatory instructions, giving families an opportunity to apply problem-solving skills and strategies. Stations allow people to choose the time they spend on an activity, setting their own pace.

While families are working at the stations, circulate among them providing encouragement and suggestions where needed. Make observations of points to be discussed later. Be careful not to give away the answers. It's OK to give families clues or ask questions that help them solve the problems. When you close this period of the session, explain why these activities were chosen, how they fit into the curriculum, and the mathematics involved.

Lesson plans

Sample lesson plans that help you pick and choose how to organize your class are provided on pages 157–158. At the end of each class, consider providing people with paper to write their comments about the evening. If you tell them they don't need to write their names on the comment sheets, the comments are more likely to be candid. This will help you revise and refine your next class.

Handouts

Be sure to give the families copies of all the activities they've done in class and reminders about using them at home during the week. At your next class meeting, discuss the activities the families did at home. Encourage parents and children to keep journals of their experiences as they explore mathematical ideas.

THE ACTIVITIES

Topics

The activities in this book are arranged by mathematical topic. In selecting the activities you want to use, you may decide to teach from one topic each class or from several topics each time. Many of the activities integrate several math strands. All the activities in the book have been chosen because they promote problem solving and mathematical reasoning. They involve the use of concrete materials, and can be enjoyed again and again. Most of the activities can be interesting to a wide age span. Parents can adjust the activities for both older and younger children.

Directions

Spend some time at the beginning introducing each new activity with clear directions. You may want to stop the activity about halfway through to talk about strategies families are using, patterns they see, or predictions they want to make. Your goal is to guide families to their own solutions and strategies, stressing the importance of the process rather than simply getting the right answer. Point out the mathematical skills involved and how the activity relates to the school curriculum.

One of the greatest gifts you can give parents and children is the understanding that getting the answer is not as important as understanding how to get there. This message should be repeated often throughout the course.

Closure

As you close the activity, repeat its connection to the curriculum and the math skills involved. You may want the families to brainstorm the skills and strategies they have used to solve the problem; you may want to bring closure to the problem by having the whole group work all the way through a solution process; or you may decide to let the families work it through on their own and report the following week on their progress.

No matter how you organize your class, it is important always to provide written activity instructions for the families to refer to at home. You also may want to include copies of game boards, special dice, etc., in the materials you send home.

"FAMILY MATH *is a simple program with profound results.*"

–Dr. Eugene Cota-Robles

IN OTHER LANGUAGES

Non-English-speaking families

FAMILY MATH can be an excellent vehicle to reach families who do not speak English. It provides a comfortable way for them to become involved with school and to learn the importance of mathematics in this culture. If the instructor does not speak the second language, a translator will be needed, preferably one who is interested in or understands the activities. Sometimes other bilingual parents may be able to provide translations. Be sure to allow time for these conversational translations to be completed before moving on too quickly.

It is important to provide translations of the handouts, if possible, and to make sure all questions are answered. Check with the translator frequently. There may be many questions, since this way of approaching mathematics may differ from the traditional teaching of many countries. Your explanations will be important in encouraging the families to return each week.

Send home recruitment notices in the language of the parents you want to attract. Word-of-mouth and sample activities may also help. In any case, you will find this kind of class among the most satisfying to you personally, as you watch parents and children enjoy mathematics together.

FAMILY MATH ACROSS THE GRADES

Special note for junior high school or middle school classes: FAMILY MATH is available for families with children in grades five through nine in *FAMILY MATH—The Middle School Years, Algebraic Reasoning and Number Sense*. For families with children in grades PreK through three, *FAMILY MATH for Young Children* classes are also popular.

INTRODUCING CAREERS

Careers

An important part of FAMILY MATH classes is providing information about careers and including role model speakers who visit the class. When we developed our first FAMILY MATH class, we wanted families to meet with men and women working in a range of fields (from the skilled trades to business to research scientists) to demystify some of the questions about future career choices in math-related fields. We made sure that our role models understood our emphasis on the importance of mathematics to future options, so they were able to tell personal stories about their experiences in learning and doing mathematics. We also tried to include role models who worked in nontraditional fields, such as women who were plumbers and men who were nurses.

Role models

You can invite one person or a panel of three or four role models. To find role models, ask members of the class, your friends, trade schools, university career center, or organizations at the local college such as the Society of Women Engineers.

When the role model activity begins, ask each of the role models to talk for about five minutes about the work they do or the field they are studying and how they chose that field. Suggest that they talk about:
- how parents and teachers influenced their choices
- who their mentors were and what kinds of encouragement were provided
- any obstacles they encountered along the way and how they persisted in spite of them

Hold all questions until each role model has spoken. Be prepared with a math activity in the rare event that the questions run out before the class is over.

EVALUATION

It might seem strange to think about evaluation in such an informal setting, but there are a number of benefits in knowing what you've accomplished and how you might improve the class. Here are our thoughts on evaluation.

Why should I evaluate?
- For my own information, to help me improve the class
- For administrators, school board, community interest
- For potential funders, who like to know what they're supporting

What could I evaluate?
- Who comes? Why did they come? Who doesn't come or doesn't come back? Why didn't they?
- How is the class going? Are people enjoying it? Are my presentations and explanations clear? Am I making the connections that people need?
- Are they more positive about math, about helping their children with math, and about their child's math education? Does everyone feel more confident?
- What are people doing between classes? Do they use any of the activities? Do they find more math activities to try?
- What are some long-term outcomes of the class? Are parents more involved with the school? Are they getting more involved with their children's math education?
- Are parents learning more math themselves? Are children more interested in math?

How can I evaluate?
- Informal comments—begin the class with introductions and a question, like: "How did you hear about the class?" "Why did you come?" "Did you try any activities from last week?"
- Journals—ask parents to keep notes of the classes in journals where they also record math activities and reflections they have between classes. Borrow these journals to read at the end of the sessions.
- Checklists—list the activities you've presented by name, and ask parents to rate them (liked it, didn't like it; tried it at home, plan to try it sometime, don't plan to try).
- Open-ended written comments made at the end of each class.
- Follow-up strategies—talk to children's teachers, talk to children, have a follow-up meeting of parents.

Any other advice?
- Gather only data that you plan to use, ask questions for a reason, and try to act on people's recommendations if you agree that they're sound.
- Listening carefully is one of the best evaluation techniques available!

BEST OF LUCK!

We hope that this section on how to set up a FAMILY MATH class has been helpful to you. We have tried to give you enough information to proceed without overwhelming you with ideas. We'd love to hear how your class went. If you have questions or comments, please contact FAMILY MATH at the Lawrence Hall of Science, University of California, Berkeley, CA 94720-5200, and (510) 642-1823. Our web site address is www.lawrencehallofscience.org/equals and our email address is equals@uclink.berkeley.edu.

FAMILY MATH
PLANNING CHECK SHEET
Things to do before class:

WHEN	WHAT
1 or 2 months before class	❑ Decide on time and place ❑ Decide on grade levels ❑ Make space arrangements with school, community center or church
About 6 weeks before class	❑ Begin recruiting *(Much earlier than 8 weeks and people may forget, and later than 3 weeks doesn't leave enough time for parents to plan)*
1 or 2 weeks before each class	❑ Finalize class curriculum ❑ Begin to gather needed materials ❑ Prepare masters for handouts ❑ Arrange child care if needed
About 2 weeks before the role model or panel	❑ Select date and recruit role models for career activity
1 week before class	❑ Run off handouts for week 1 *(guess on enrollment)* ❑ Double-check room availability ❑ Send home reminder notices to those who signed up
2 hours before class	❑ Set up openers and sign-in sheet. ❑ Arrange furniture the way you like it ❑ Set up refreshments.
When class begins	❑ Relax, it's going to be wonderful!

SAMPLE LESSON PLAN
Grades K–3, Lesson 1

TIME	ACTIVITY	HANDOUTS/MATERIALS
7:00–7:10 7:10–7:20	**Opener and stations** **Preview the evening** **Welcome** Introductions Parents say who they are and introduce their children. Which station did you visit first?	• related materials • welcome sign • sign-in sheet
7:20–7:35	**Bean Boxes** • Ask families if they have ever had to divide something evenly. • Model dividing the beans, and ask families to help you complete the activity.	• pencils, paper • beans • Bean Boxes boards
7:35–7:40	**Name Game** • What types of letters does your name have the most of? • Which letters have more open curves: upper case or lower case letters? • Ask other questions you can think of as the activity progresses.	• Name Game handout • pencils, paper
7:40–7:55 board	**Balloon Ride Revisited** • Give Instructions • Model first 3 moves	• toothpicks • markers • Balloon Ride Revisited game
7:55–8:10	**Snacks** refreshments and stations	
8:10–8:20	**Snail Races** • Ask families what they know about probability and fair dice. • Talk about the probability (chance) of rolling the different numbers on a single die. • Model the game with a partner to the third or fourth turn. Then let the families get started on their game.	• markers • dice • pencils • Snails Races game boards
8:20–8:25	**Discussion** • Raise your hand if your snail won. • Which snails do you think will win more than others? Why?	
8:25–8:30	**Session ends** • Distribute activity handouts • Homework • Comment Cards	• homework handout and related materials, activity handouts

Note: This schedule is meant to be flexible. The goal is to give you and the families enough time to get into the mathematics and discuss the strategies.

SAMPLE LESSON PLAN
Grades 4–6, Lesson 2

TIME	ACTIVITY	HANDOUTS / MATERIALS
7:00–7:10	**Opener and stations**	• stations and related materials • welcome sign • sign-in sheet
7:10–7:20	**Welcome** **Preview** the evening **Introductions** Children introduce their parents, or they tell one fun or interesting thing they do with their families.	
7:20–7:40	**Grampa's Coins** • Discuss counting money • Model one of the combination sets • Pose a question about money or saving it such as "How old were you when you learned to count money?"	• real pennies, dimes, quarters; or play coins • calculators
7:40–7:45	**Discussion** • Ask families about the different combinations that they made. • Have them share their results with another family; then ask two or three families to share with the larger group.	
7:45–8:00	**Refreshments**	
8:00–8:15	**Loopy Lou II** Discuss • number patterns • adding and subtracting This is a way to practice adding in your head.	•spinners •scratch paper •calculators (optional) •pencils
8:15–8:20	**Discussion** • What patterns did you notice? • Were there numbers that seemed like they would never reappear in the sequence? • Which numbers came up rather quickly within a sequence? • What did you notice in the patterns?	
8:20–8:30	**Closure** • Have families take home Loopy Lou II and ask them to find other patterns. • Give out activity handouts. • Comments	• evaluation sheets • Loopy Lou II and other activity handouts

Note: You may want to go over homework activities later in the class series. For some classes, you may want to wait to discuss the activities all at one time.

"Mathematics rightly viewed possesses not only truth, but supreme beauty."

–Betrand Russell

What Mathematics is Taught in Grades K-6?

This section describes *in general* what your child should learn and understand at each grade level. It is based on guidelines created by the National Council of Teachers of Mathematics (NCTM). Ask your child's teacher for information regarding local and state standards.

This list is meant to guide. Keep in mind that all children are different and develop at their own rate. There are no hard rules about the age that most students should learn most topics. Keep in mind that we should help children exceed rather than merely meet minimum proficiencies or standards.

We mention the use of calculators throughout this section. We feel strongly that calculators are critical to solving long and tedious calculations, and that they allow children to grasp more sophisticated concepts at a younger age.

MATHEMATICS GENERALLY COVERED IN KINDERGARTEN

Applications
- Talk about mathematics in their daily lives.
- Use tools and strategies to make decisions about how to solve a problem.
- Explain their thinking and reasoning in words, or using illustrations or objects.

Numbers
- Learn to estimate.
- Count objects, up to about 30.
- Count objects to match a number.
- Compare two sets of objects (up to 10 objects each) and determine which sets are equal to (=), more than (>), or less than (<) the other.
- Recognize numerals at least up to 30.
- Write numerals, 0 through 9.
- Understand simple addition and subtraction (for two numbers, each less than ten).

Measurement
- Estimate and compare various measurements: taller or shorter, longer or shorter, larger or smaller, heavier or lighter.

Geometry
- Recognize and sort colors.
- Recognize and sort simple shapes (circles, squares, rectangles, triangles).
- Recognize and name simple solid shapes (cube, sphere, cone).
- Build or illustrate simple shapes such as triangles, squares, cylinders.
- Give and follow directions by location (above, below, under, over).

Patterns
- Recognize simple patterns, continuing them, and making up new patterns such as ab-ab-ab, and so on.

Statistics
- Collect information about everyday things, such as birthdays, pets, food, and so on. Make simple graphs using the information you collect, and talk about them.

What Mathematics is Taught in Grades K-6?

MATHEMATICS GENERALLY COVERED IN FIRST GRADE

Applications
- Talk about mathematics used or observed in daily living.
- Learn strategies such as using objects or drawing diagrams to solve problems.
- Create problems using addition and subtraction to solve.
- Practice estimation skills.

Arithmetic and Numbers
- Count, read, and write, whole numbers through at least 100.
- Practice counting by twos, fives, and tens.
- Use ordinal numbers, such as first, second, tenth, and so on.
- Learn basic addition and subtraction facts up to $9 + 9 = 18$ and $18 - 9 = 9$.
- Find the sum of three one-digit numbers ($1 + 3 + 2 = \_\_$).
- Understand and represent equivalent forms of the same number (to 20). For example, the number 6 can be represented as $1 + 5$, $2 + 4$, $3 + 3$, $9 - 3$, $8 - 2$, and $7 - 1$.
- Develop understanding of place value (tens and ones) using manipulatives such as base ten blocks, Cuisenaire rods, abaci, money.
- Develop the concept of fraction values including halves, thirds, and fourths.
- Identify the value of coins; develop an understanding of equivalent combinations and values.

Measurement
- Tell time to the hour or half-hour.
- Recognize and use calendars, days of the week, months.
- Estimate lengths and measure with nonstandard units, such as how many handspans across the table.
- Understand the relative values of pennies, nickels, dimes.

Geometry and Patterns
- Work with shapes, such as triangles, circles, squares, rectangles.
- Recognize, repeat, and make up patterns of numbers and shapes.
- Classify plane (flat) and solid (not flat) objects by common characteristics.
- Arrange and describe objects in space by proximity (next to, far, near, in front of).

Probability and Statistics
- Make and interpret simple graphs of everyday things, such as color preferences, number of brothers and sisters, and so on.
- Compare data (largest, smallest, most often, least often) by using bar graphs, tally charts, and picture graphs.

Algebra
- Solve number sentences from problems that express relationships involving addition and subtraction. For example, Ryan had some books. He bought three more books. Now he has 8 books. How many did he have to start with?
- Understand the meaning of the symbols +, -, and =.

MATHEMATICS GENERALLY COVERED IN SECOND GRADE

Applications
- Talk about mathematics used in daily life.
- Create and solve word problems in measurement, geometry, probability, and statistics, as well as arithmetic.
- Practice strategies for solving problems, such as drawing diagrams, organized guessing, and paraphrasing.

Arithmetic and Numbers
- Practice estimation skills.
- Count, read, and write numbers through at least 1,000 and identify the place value for each digit.
- Identify odd and even numbers.
- Understand, identify, and compare fractions.
- Understand and use the symbols for greater than (>) and less than (<).
- Know addition and subtraction facts through $9 + 9 = 18$ and $18 - 9 = 9$.
- Estimate answers to other addition and subtraction problems.
- Practice addition and subtraction with and without regrouping (carrying), in problems such as:

```
  24      67      57      47
 + 2     + 8     - 2     - 8
  26      75      55      39
```

- Add columns of numbers, such as:

```
   1
   5
   8
 + 3
```

- Explore uses of a calculator.
- Use repeated addition to do multiplication.
- Use repeated subtraction to do division.

Geometry
- Find congruent shapes (objects that have the same size and shape).
- Identify squares, rectangles, circles, and other polygons.
- Recognize lines of symmetry.
- Read and draw simple maps.

Measurement
- Practice estimation of measurements.
- Measure with nonstandard units-how many toothpicks long is the table?
- Use standard units such as inches or centimeters.
- Tell time and understand the relationship of minutes in an hour, hours in a day, etc.
- Know days of the week and months, and use the calendar to find dates.
- Make change with coins and bills.
- Solve money problems with manipulatives.

Statistics
- Make and interpret simple graphs.
- Identify features of data sets, such as which event happens the most frequently (mode) and the difference between the highest and lowest data points (range). For example, if the lowest score on the spelling test was 6 correct and the highest score was 9, then the range is 6 to 9.
- Keep track of numerical data.
- Ask and answer simple questions related to numerical data.

Patterns
- Work with patterns of numbers, shapes, colors, sounds, and so on; including adding to existing patterns, completing missing sections, and making up new patterns.
- Recognize, describe, and extend linear patterns to the next term. For example, the number of feet on one canary is 2, the number of feet on two canaries is 4, the number of feet on three canaries is 6 and so on.

Algebra and Functions
- Use the commutative and associative rules to simplify mental calculation and to check results.
  For example; $12 + 11 = 11 + 12$, or $(12 + 11) + 3 = 12 + (11 + 3)$.
- Solve addition and subtraction problems using charts, picture graphs, or number sentences.

What Mathematics is Taught in Grades K-6?

MATHEMATICS GENERALLY COVERED IN THIRD GRADE

Applications
- Talk about mathematics in their lives.
- Create, analyze, and solve word problems using a variety of methods.
- Use what is applied in a particular problem as a generalization for other problems.

Numbers
- Practice estimation skills with all problems.
- Identify the place value for each digit for numbers up to 10,000.
- Name and compare fractions such as $\frac{1}{2}$ and $\frac{6}{12}$.
- Add and subtract simple fractions. For example, determine that $\frac{1}{8} + \frac{3}{8}$ is the same as $\frac{1}{2}$.
- Identify fractions of a whole number, such as $\frac{1}{2}$ of 12 is 6.
- Use money to explore concepts of decimal numbers, such as tenths and hundredths.
- Use the signs for greater than (>) and less than (<).

Arithmetic
- Use calculators effectively, and apply this knowledge to solving problems.
- Continue to work with patterns, including those found on addition and multiplication charts.
- Understand and use inverse relationships of addition and subtraction, and multiplication and division to compute and check results.
- Solve larger and more complicated addition and subtraction problems, such as:

$$\begin{array}{r} 2{,}887 \\ +\,6{,}395 \\ \hline \end{array} \qquad \begin{array}{r} 8{,}356 \\ -\,1{,}967 \\ \hline \end{array}$$

- Learn multiplication and division facts through $9 \times 9 = 81$ and $81 \div 9 = 9$.
- Begin to learn multiplication and division of two- and three-digit numbers by a single-digit number, such as:

$$\begin{array}{r} 16 \\ \times\,6 \\ \hline \end{array} \qquad \begin{array}{r} 106 \\ \times\,7 \\ \hline \end{array} \qquad 4\overline{)64}$$

- Learn about remainders in problems such as: $29 \div 7 = 4\,R1$.

Measurement and Geometry
- Estimate before measuring length, volume, area, perimeter.
- Understand and apply various standard measures such as centimeters, decimeters, meters, inches, feet, yards.
- Measure perimeter (using units of length), area (using square units: square centimeters, square inches, square feet), weight (kilograms, pounds), volume and capacity (liters, cubic centimeters, gallons), temperature (Celsius, Fahrenheit).
- Tell time to the nearest minute.
- Use money to develop understanding of decimals.
- Use calendars.
- Recognize and name shapes such as squares, rectangles, trapezoids, triangles, circles; and three-dimensional objects such as cubes and cylinders.
- Identify attributes of quadrilaterals and triangles. For example, a square has four equal sides and right angles; an isosceles triangle has three sides, of which two are equal.
- Identify congruent shapes (objects that are the same size and shape).
- Recognize lines of symmetry, reflections (mirror images) and translations (movements to a different position) of figures.
- Read and draw simple maps, using coordinates.
- Learn about parallel (| |) and perpendicular ($\perp$) lines.

Statistics
- Record the possible outcomes for a simple event, and systematically keep track of the outcomes when the event is repeated many times (for example, tossing a coin).
- Summarize and display the results of probability experiments in a clear way (for example, using a bar graph, a line plot, or the like).
- Identify whether an event or occurrence is certain, likely, unlikely, or impossible.
- Use results of probability experiments to predict future events.

Algebra and Functions
- Select appropriate symbols (+, −, ÷, x) to make an expression true. For example: 5 x 3 = 15.
- Represent simple functional relationships. For example, finding the total cost of several identical items, given the cost per unit.
- Extend and recognize linear patterns by their rules. For example, the number of legs on a given number of birds may be calculated by counting by twos or by multiplying the number of birds by 2.
- Know and understand that, if two numbers are equal, then the results of adding a third number to each will be equal as well. For example, if 5 = (1 + 4), then 5 + 9 = (1 + 4) + 9.
- Know and understand that, if two numbers are equal, then the results of multiplying each by a third number will be equal as well. For example, if 5 = (1 + 4), then 5 x 9 = (1 + 4) x 9.

MATHEMATICS GENERALLY COVERED
IN FOURTH GRADE

Applications
- Talk about uses of mathematics in students' lives and in their futures.
- Create, analyze, and solve word problems in all of the concept areas.
- Use a variety of problem-solving strategies to solve problems with multiple steps.
- Apply strategies from simpler problems to more complex problems.
- Develop formal and informal mathematical vocabulary.

Numbers and Operations
- Use calculators with some proficiency for all operations.
- Read and write numbers to the millions, and know how to round to whole numbers.
- Learn about special numbers, such as primes, factors, multiples, and square numbers.
- Recognize equivalent fractions, such as

$$\frac{1}{2} = \frac{2}{4}$$

- Find fractions of whole numbers, such as

$$\frac{1}{4} \text{ of } 20 = 5$$

- Maintain and extend work with the operations of addition, subtraction, multiplication, and division of whole numbers.
- Add and subtract simple decimal numbers.
- Learn about simple percents such as 10%, 50%, and 100%.
- Use concepts of negative numbers. For example, on a number line or reading temperature.
- Identify the relative position of positive fractions, positive mixed numbers, and positive decimals to two decimal places on a number line.

Geometry
- Find patterns in geometric shapes.
- Recognize right angles; explore terminology of other angles.
- Identify various types of symmetry.
- Read and draw simple maps, using coordinates.
- Understand terminology and uses of coordinate grid.
- Identify parallel (| |) and perpendicular (⊥) lines.
- Explore how different shapes fill a flat surface (tiling a floor, for example).

Measurement
- Estimate before measuring.
- Use standard units to measure length, area, volume, weight, temperature.
- Make simple scale drawings.

Statistics, Data Analysis, and Probability
- Use sampling techniques to collect information or conduct a survey.
- Discuss uses and meanings of statistics, such as how to make a fair survey, how to show the most information, and how to find averages.
- Make, read, and interpret graphs.
- Perform simple probability experiments and discuss their results.

Algebra
- Demonstrate understanding of patterns, relations, and functions using words, tables, and graphs to describe, extend, and analyze them.
- Use algebraic symbols to represent and analyze mathematical situations.
- Identify, describe, and compare situations with constant or varying rates of change; and compare them.
- Use and interpret formulas (area = length x width or $A = lw$) to compare quantities.
- Understand the use of parentheses in the order of operations. For example, distinguish $2 + (3 \times 7) - 5 = 2 + (21) - 5 = 18$ which is different from $(2 + 3) \times (7 - 5) = (5) \times (2) = 10$.

MATHEMATICS GENERALLY COVERED IN FIFTH GRADE

Applications
- Work in groups to solve complex problems.
- Create, analyze, and solve word problems.
- Use a variety of strategies to solve problems with multiple steps.
- Compute a given percent of a whole number.
- Use calculators effectively for appropriate problems.
- Develop mathematical vocabulary, and apply it appropriately.

Numbers and Operations
- Expand understanding and use of special numbers such as primes, composite numbers, square and cubic numbers, common divisors, and common multiples.
- Interpret percents as a part of a hundred; find decimal and percent equivalents for common fractions and understand why they represent the same value.
- Identify decimals, fractions, mixed numbers, and positive and negative integers.
- Add and subtract positive and negative numbers, and verify the reasonableness of the results.
- Demonstrate proficiency with division, including division with positive decimals and long division with multidigit divisors.
- Understand how prime factors are determined and write the multiples as the product of their prime factors by using exponents For example $24 = 2 \times 2 \times 2 \times 3 = 2^3 \times 3$.
- Solve simple problems involving the addition and subtraction of fractions and mixed numbers.

Measurement and Geometry
- Understand and apply the formula for the area of a triangle and of a parallelogram by comparing each with the formula for the area of a rectangle. That is, two of the same triangles make a parallelogram with twice the area.
- Measure, identify, and draw angles, perpendicular and parallel lines, rectangles, and triangles by using appropriate tools, such as straight-edge, ruler, compass, protractor, drawing software.
- Know that the sum of the angles of any triangle is 180° and the sum of the angles of any quadrilateral is 360°, and use this information to solve problems.

Statistics
- Perform and report on a variety of probability experiments.
- Collect and organize data.
- Understand and determine the mean, median, and mode of a specific set of data.
- Display data in graphic form, such as bar, picture, circle, and line graphs.
- Identify ordered pairs of data from a graph and interpret their meaning.

Algebra and Functions
- Develop some understanding of ratio and proportion.
- Use a letter to represent an unknown number; write and evaluate simple algebraic expressions in one variable by substitution.
- Identify and graph ordered pairs in the four quadrants of the coordinate plane.
- Solve problems involving linear functions with integer values, write the equation, and graph the resulting ordered pairs of integers on a coordinate plane.

What Mathematics is Taught in Grades K–6?

MATHEMATICS GENERALLY COVERED IN SIXTH GRADE

Applications
- Make connections between mathematics and daily life.
- Create, analyze, and solve word problems in all of the concept areas.
- Apply a variety of strategies to solve problems with multiple steps.
- Work with others to solve complex problems.
- Use calculators effectively for problem solving.
- Develop mathematical vocabulary, and apply it in meaningful contexts.

Numbers and Operations
- Practice rounding and estimation skills with all problems.
- Continue to increase understanding of fraction relationships like:

    Comparisons, such as $\frac{2}{3} > \frac{1}{2}$

    Equivalence, such as $\frac{2}{3} = \frac{4}{6}$

    Reducing, such as $\frac{10}{20} = \frac{1}{2}$

    Relating mixed numbers and improper fractions, such as $2\frac{1}{3} = \frac{7}{3}$

- Continue to develop skills in adding, subtracting, multiplying, dividing fractions and decimal numbers.
- Compare and order positive and negative fractions, decimals, and mixed numbers; and place them on a number line.
- Interpret and use ratios in different contexts. to show the relative sizes of two quantities, using appropriate notations (a/b, a to b, $a{:}b$). For example, batting averages, miles per hour.
- Use proportions to solve problems. For example, determine the value of N if $\frac{4}{7} = \frac{N}{21}$; find the length of a side of a polygon similar to a known polygon.
- Calculate percentages of quantities; and solve problems involving discounts at sales, interest earned, and tips.
- Determine the least common multiple and the greatest common divisor of whole numbers; use them to solve problems with fractions. For example, to find a common denominator when adding two fractions.
- Solve addition, subtraction, multiplication, and division problems.
- Explore scientific notation, such as $3 \times 10^8 = 300{,}000{,}000$.

Geometry
- Deepen understanding of the measurement of plane and solid shapes, and use this understanding to solve problems.
- Identify the properties of two-dimensional figures.
- Understand the concept of a constant such as ϖ; know the formulas for the circumference and area of a circle.
- Understand and use the concept of parallel (||) and perpendicular (⊥) lines.
- Understand the properties of complementary and supplementary angles, and apply them to solve problems involving an unknown angle.
- Measure and draw various angles using appropriate tools, such as straight-edge, ruler, compass, protractor, and drawing software.
- Develop understanding of symmetry, reflections, and translations of figures.
- Draw constructions such as equal line segments or perpendicular bisectors.
- Understand coordinate graphing.
- Know that the sum of the angles of any triangle is 180° and the sum of the angles of any quadrilateral is 360°, and use this information to solve problems.

Measurement
- Continue to use hands-on tools of measurement. Estimate and measure length, area, volume and capacity, and temperature.
- Convert one unit of measure to another. For example, from feet to miles, from centimeters to inches.

Statistics
- Conduct a variety of probability experiments.
- Compute and analyze statistical measurements using range, median, and mode of data sets.
- Display data in graphic form, such as bar, picture, circle, line, and other graphs.
- Use data to estimate the probability of future events.
- Understand the use of outliers and their effects on measures of central tendency.
- Identify various ways of selecting. For example, surveys, sampling, and census.
- Question and determine the validity of statistical claims (identify bias).

Algebra and Functions
- Write and solve one-step linear equations with one variable.
- Understand and apply algebraic order of operations.
- Understand the concept of *rate;* and apply it to solve problems involving average speed, distance, and time.
- Demonstrate an understanding that rate is a measure of one quantity per unit value of another quantity.

BOX MATH A (3, 5, 7) and C (3, 5, 7, 9) Number and Operation Cards

BOX MATH B (4, 6, 8) and D (2, 4, 6, 8) Number and Operation Cards

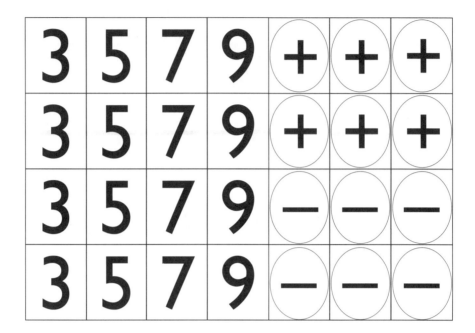

BOX MATH E F — MULTIPLICATION & DIVISION

BOX MATH E (2, 4, 6) Number and Operation Cards

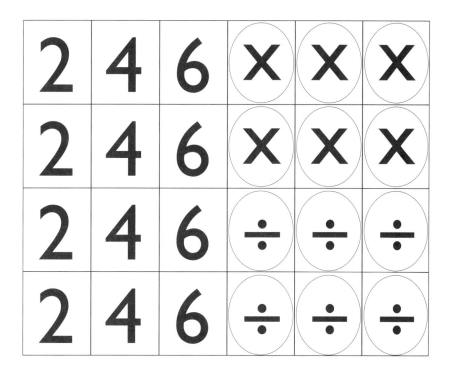

BOX MATH F (3, 6, 9) Number and Operation Cards

BOX MATH

G H

MULTIPLICATION & DIVISON

BOX MATH G (2, 4, 8) and H (2, 4, 5) Number and Operation Cards

2	4	5	8	×	×	×
2	4	5	8	×	×	×
2	4	5	8	÷	÷	÷
2	4	5	8	÷	÷	÷

FAMILY MATH II: ACHIEVING SUCCESS IN MATHEMATICS

SPINNER

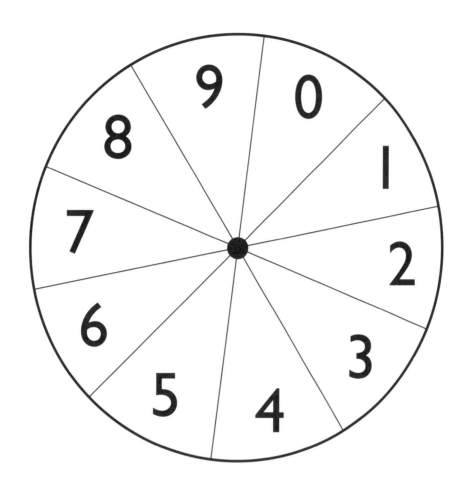

HOW TO MAKE YOUR SPINNER:
1. Make a copy of the above spinner.
2. Unfold a paper clip so that one edge is pointed straight as shown below.

3. Place the looped end of the paper clip on the center of the spinner.
4. Hold the paper clip in place with a pen or pencil.
5. Spin the paper clip by flicking the pointed end with your finger.

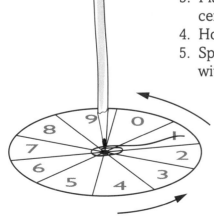

FAMILY MATH II: ACHIEVING SUCCESS IN MATHEMATICS

1 INCH GRID

2 CENTIMETER GRID

ICE CREAM CONE MATH — SCOOPS & CONES

PAPER BEADS FOR SIMONE AND BRIANNA

FAMILY MATH II: ACHIEVING SUCCESS IN MATHEMATICS

ONE-HUNDRED CHART

1	2	3	4	5	6	7	8	9	10
11	12	13	14	15	16	17	18	19	20
21	22	23	24	25	26	27	28	29	30
31	32	33	34	35	36	37	38	39	40
41	42	43	44	45	46	47	48	49	50
51	52	53	54	55	56	57	58	59	60
61	62	63	64	65	66	67	68	69	70
71	72	73	74	75	76	77	78	79	80
81	82	83	84	85	86	87	88	89	90
91	92	93	94	95	96	97	98	99	100

OTHER RESOURCES

Other EQUALS Books:

Assessment Alternatives In Mathematics. Jean Kerr Stenmark, editor.

Get It Together. Tim Erickson

FAMILY MATH. Jean Kerr Stenmark, Virginia Thompson, Ruth Cossey

FAMILY MATH for Young Children: Comparing. Grace Coates, Jean Kerr Stenmark

FAMILY MATH—The Middle School Years, Algebraic Reasoning and Number Sense. Karen Mayfield-Ingram, Virginia Thompson

For information or to order call 800.897.5036.

Other Resources

Beyond Facts and Flashcards: Exploring Math With Your Kids. Jan Mokros and TERC. Heinemann. Portsmouth, NH.

Extraordinary Play with Ordinary Things. Make-it-yourself, do-it-yourself activities that encourage your child's development. Sher, Barbara. Bob Adams Inc., Holbrook, MA.

Fear of Math: How to Get Over It and Get On with Your Life. Claudia Zaslavsky. Rutgers University Press, New Brunswick, NJ.

Math for Humans: Teaching Math through 7 Intelligences. Mark Wahl,. Livnlern Press. Langley, WA.

Math on Call: A Mathematics Handbook. Great Source Education Group. Houghton Mifflin. Wilmington, MA.

Parent Partners: Workshops to Foster School/Home/Family Partnerships. Jacqueline Barber with Lynn Barakos and Lincoln Bergman. Lawrence Hall of Science. Berkeley, CA.

The Do's and Don'ts of Parent Involvement. Katherine Kellison McLaughlin. INNERCHOICE Publishing, Anaheim, CA.

The Sizeasaurus: From Hectares to Decibels to Calories, A Witty Compendium of Measurements. Stephen Strauss. Kodansha America, Inc., New York, NY.

Song titles mentioned in *FAMILY MATH II*:

"18 Wheels on a Big Rig"
http://songsforteaching.homestead.com
   1. Scroll down to:
      Using Children's Music to Teach Content Area Subjects,
   2. Select:
      Mathematics link
   3. Scroll down to
      Miscellaneous Math Songs and select "18 Wheels on a Big Rig"

"I Caught a Fish Alive"
www.scholastic.com/smartparenting/timetogether/ages0_2/rhymereason.htm
   Scroll down to song 2, "I Caught a Fish Alive."

"Ladybugs' Picnic"
http://members.tripod.com/Tiny_Dancer/ladybug.html

Web Resources

To find out about upcoming FAMILY MATH workshops or to find a FAMILY MATH site near you, check our website at:

www.lawrencehallofscience.org/equals

We also recommend you search the internet under "family learning."

BIBLIOGRAPHY

Cecil, Nancy Lee. *The Art of Inquiry: Questioning Strategies for K-6 Classrooms.* Peguis Publishers. Winnipeg, MB. 1995.

Center for Educational Research and Innovation. *Parents as Partners in Schooling.* Copyright Clearance Center, Inc. USA. 1997.

Coates, Grace D., Jean Kerr Stenmark. *FAMILY MATH for Young Children: Comparing.* UC Printing, University of California, Berkeley. Berkeley, CA. 1997.

Gardner, Howard. *Multiple Intelligence: Theory in Practice, A Reader.* Basic Books, Harper Collins Publishers, Inc., New York, NY. 1993.

Great Source Education Group. *Math on Call.* Great Source Education Group, Inc. Wilmington, MA. 1998.
——— *Math to Know.* Great Source Education Group, Inc. Wilmington, MA. 1998.

Ma, Liping. *Knowing and teaching Elementary Mathematics. Teachers' Understanding of Fundamental Mathematics in China and the United States.* Lawrence Erlbaum Associates, Inc. Mahwah, NJ. 1999.

National Research Council. Bransford, John D., et.al. editor *How People Learn. Brain, Mind, Experience, and School.* National Academy Press, Washington, D.C. 1999.

Sher, Barbara. *Extraordinary Play with Ordinary Things.* Make-it-yourself, do-it-yourself activities that your child's encourage development. Bob Adams Inc., Holbrook, MA. 1994.

National Council of Teachers of Mathematics and National Association for the Education of Young Children. Juanita V. Copley, editor. *Mathematics in the Early Years.* The National Council of Teachers of Mathematics, Inc. Reston, VA. 1999.

Stenmark, Jean Kerr, Virginia Thompson, Ruth Cossey. *FAMILY MATH.* University of California. UC Printing, Berkeley, CA. 1986.

Wahl, Mark. *Math for Humans. Teaching Math through 7 Intelligences.* Livnlern Press. Langley, WA. 1997.

INDEX

A
A PENNY SAVED, **40**
ABOUT YOUR HEIGHT AND MORE, **32**
ACKNOWLEDGEMENTS, **VI**
Addend, 82
Addition, 40, 66, 68, 74, 76, 86, 88, 90, 108, 115
Algebra, 38, 40, 42, 48, 50, 54, 57, 58, 62
Alphabet Cards, 36
ALGEBRAIC AND LOGICAL THINKING INTRODUCTION, **38**
Area, 54, 57, 136
Average, 32

B
BALLOON RIDE REVISITED, **1**
BASEBALL CARDS, **50**
BEADED BRAIDS: INVESTIGATING PATTERNS AND RATIOS, **70**
BEAN BOXES, **84**
BIBLIOGRAPHY, **185**
Biology, 58
BIRTHDAY GRAPH, **18**
Birthday twins, 18
Body estimates, 32
BODY TRACE, **114**
BOX MATH ADDITION AND SUBTRACTION, **76**
BOX MATH MULTIPLICATION AND DIVISION, **94**
Business, 32

C
Calendar, 18, 112
CIRCULAR NIM, **46**
Circumference, 130
CLEAN YOUR PLATE, **68**
Collections, 50
Combinations, 20, 30, 34
Comparing, 50, 112
Composing numbers, 76
Conductor, 92
CONSECUTIVE SUMS, **82**
Counting, 27, 40, 43, 68, 74
CREATE A FAMILY MATH KIT FOR HOME, **14**
CREATING AN ENVIRONMENT FOR LEARNING, **2**
Curve, 118

D
Decimals, 94
Decomposing numbers, 76
Denominator, 100
Dice, 22, 24, 28
Direction, 28
Discrete math, 20, 27, 30, 34
Division, 66, 84, 88, 94
DOT TO DOT, **120**
Double birthday, 18
DOUBLING BEAN BOXES, **86**

E
Economics, 32, 58
Engineering, 58
Estimation, 32, 74, 114, 130, 136
Exponential Growth, 27
Exponents, 100

F
Factor, 30, 58, 84, 86
Factorial, 31
FAMILY GARDEN, **100**
FAMILY GARDEN GAME, **104**
FAMILY MATH ON THE GO, **74**
Fractions, 27, 100, 104, 136
Frobenius, 92
Function, 54, 57

G
Game Theory, 1
General case, 44
Geology, 58
GEOMETRY INTRODUCTION, **106**
Geometry, 106, 120, 122, 126, 128, 130, 132
GRAMPA'S COINS, **58**
GRANDMOTHERS, **27**
Graphs, 116

H
HOW IS MY CHILD DOING IN MATH?, **6**

I
ICE CREAM CONE MATH, **30**
Intersections, 108, 120, 132
IT ALL ADDS UP!, **115**

L
Latin Square, 62
LICENSE PLATE EXPLORATIONS, **34**
Line segment, 108, 110, 132
LINE SEGMENTS AND INTERSECTIONS, **110**
LINE SYMMETRY, **122**
Lines, 110, 118, 120, 132
Logic, 1, 38, 44, 46, 47, 50, 58, 62, 76, 84, 86, 92, 126
LOOPY LOU I, **88**
LOOPY LOU II, **90**
LOST IN SPACE, **28**

M
MANY FACES OF A CUBE I, **126**
MANY FACES OF A CUBE II, **128**
Mathematical Reasoning, 27
Matrix, 62
MASTERS FOR CHARTS, GRAPHS, SPINNERS, **174**
MATTING PICTURES, **136**
ME AND MY SHADOW, **112**
MEASUREMENT AND ESTIMATION, **130**
Measurement, 32, 112, 114, 130, 136
Mental math, 1, 76, 92, 94, 108, 115
Money, 40
Moors, 54, 57
Multiple, 58, 84, 86, 92
Multiplication, 40, 66, 74, 84, 88, 94, 115

N
NAME GAME, **118**
Nim, 1, 44, 46, 47
NO TWIN NIM, **44**
Number line, 116
Numberline Rectangles, 60
NUMBER SENSE INTRODUCTION, **64**
Number sense, 28, 58, 64, 66, 70, 82, 84, 86, 88, 90, 92, 114
Numerator, 100

O

ODD AND EVEN ENDED NIM, **47**
One-to-one correspondence, 28
Operations, 76, 94
ORGANIZING A FAMILY MATH CLASS SERIES, **142**
Organizing information, 32, 42, 48, 92, 130
Outcomes, 24

P

PATHS AND PONDS, **54**
PATIOS AND PATHS, **57**
Pattern, 42, 48, 54, 57, 62, 70, 76, 88, 90, 92, 94
Percent, 32
Perimeter, 54, 57
Permutations, 20, 30, 34
Physics, 58
Place value, 76, 90, 94
Plane, 132
Points, 120
Predicting, 42, 48
Prediction, 24
PREFACE, **V**
Prime number, 58
PROBABILITY AND STATISTICS INTRODUCTION, **16**
Probability, 18, 20, 22, 24, 28, 32, 34, 104
Proportion, 58

Q

QUESTIONS THAT PROMOTE MATHEMATICAL THINKING, **4**

R

Ratio, 32, 70
Reasoning skills, 47
Regrouping, 115
RESOURCES, **183**
ROYAL FAMILY PUZZLE, **62**

S

SAMPLE ACTIVITY PAGE, **12**
SANDWICHES, **20**
Sequence, 82, 116
Sets, 118
SIDEWALK MATH, USING SIDEWALK CHALK INTRODUCTION, **108**
SNAIL RACES, **22**
Social sciences, 32
SPAGHETTI GEOMETRY, **132**
Spatial reasoning, 26, 108, 122, 126, 128
Speed, 32
Square numbers, 48, 58
STAIRWAY TO SEVEN, **42**
Statistics, 16, 20, 22, 32, 34
Strategy, 1, 44, 46, 47
Subtraction, 66, 68, 76
Surface area, 126, 128
Symmetry Cards, 120
Symmetry, 120, 136

T

Tessellation, 57
THIS IS SIX, **66**
Time, 112
TWO COIN PROBLEM, **92**

U

UP THE DOWN STAIRCASE, **48**

V

VALUE LINE, **116**
Variables, 44
Venn Diagram, 118
Volume, 128

W

WHAT MATHEMATICS IS TAUGHT IN GRADES K–6?, **160**
WORKING WITH YOUR CHILD'S TEACHER, **9**

Z

ZOO GAME, **24**

Notes

Notes

Notes

Notes

Notes